Science

Jean Martin Andy Cooke
Sam Ellis

Core 8

CAMBRIDGE UNIVERSITY PRESS
Cambridge, New York, Melbourne, Madrid, Cape Town, Singapore, São Paulo, Delhi

Cambridge University Press
The Edinburgh Building, Cambridge CB2 8RU, UK

www.cambridge.org
Information on this title: www.cambridge.org/9780521725699

First published 2008

Book printed in the United Kingdom at the University Press, Cambridge

A catalogue record for this publication is available from the British Library

ISBN 978-0-521-72569-9 paperback with CD-ROM

Contents

Physics (continued)

Introduction

Take advantage of the CD

Cambridge Essentials Science comes with a CD in the back. This contains the entire book as an interactive PDF file, which you can read on your computer using free Acrobat Reader software from Adobe (http://www.adobe.com/products/acrobat/readstep2.html). As well as the material you can see in the book, the PDF file gives you extras when you click on the buttons you will see on most pages; see the inside front cover for a brief explanation of these.

To use the CD, simply insert it into the CD or DVD drive of your computer. If you are using Microsoft Windows, you will be prompted to install the contents of the CD to your hard drive. Installing will make it easier to use the PDF file, because the installer creates an icon on your desktop that launches the PDF directly. However, it will run just as well directly from the CD.

If you are using an Apple Mac computer or any other operating system, or you simply want to copy the disc onto your hard disc yourself, this is easily done. Just open the disc contents in your file manager (for Apple Macs, double click on the CD icon on your desktop), select all the files and folders and copy them wherever you want.

Take advantage of the web

Cambridge Essentials Science lets you go directly from your book to web-based activities on our website, including animations, exercises, investigations and quizzes. Access to these materials is free to all users of the book.

There are three kinds of activity, each linked to from a different place within each unit.

- **Scientific enquiry:** these buttons appear at the start and end of each unit. The activities in this section allow you to develop skills related to scientific enquiry, including experiments that would be hard to carry out in the classroom.
- **Check your progress:** these buttons come half-way through each unit. They let you check how well you have understood the unit so far.
- **Review your work:** these buttons come at the end of each unit. They let you show that you have understood the unit, or let you find areas where you need more work.

The *Teacher Material* CD-ROM for *Cambridge Essentials Science* contains enhanced interactive PDFs. As well as all the features of the pupil PDF, teachers also have links to the *Essentials Science* Planner – a new website with a full lesson planning tool, including example lesson plans, worksheets, practicals, assessment materials and guidance. The e-learning materials are also fully integrated into the Planner, letting you see the animations in context and alongside all the other materials.

Scientific enquiry

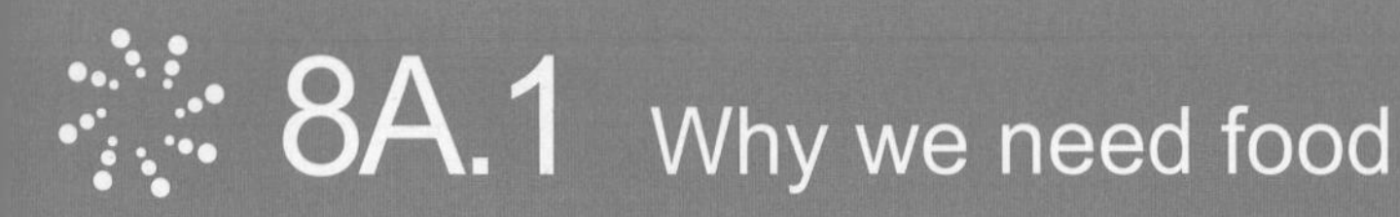

8A.1 Why we need food

You should already know | Outcomes | Keywords

Why we need food

We need food to survive.

A healthy person who stops eating will live for about 40 days.

Food gives us the raw materials that our bodies use…

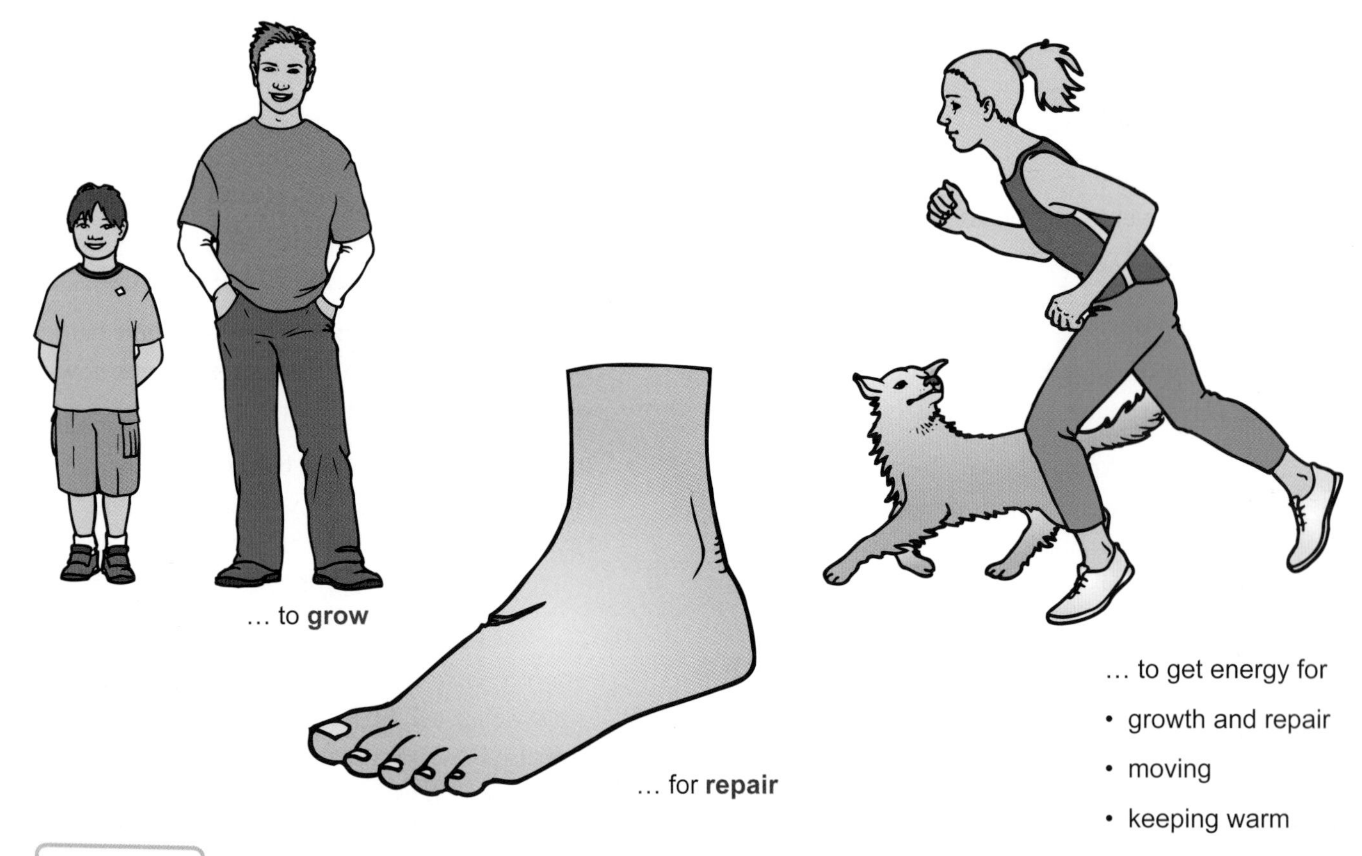

Question 1

We call the food substances that our cells use **nutrients**.

They are

- **proteins** for making new cells
- **carbohydrates** and **fats** for energy
- small amounts of **vitamins** and **minerals**.

Question 2

You are what you eat

You need protein to make new cells. So proteins are particularly important at times when you are growing quickly.

Protein foods

Question 3

Taribo has a healthy diet.

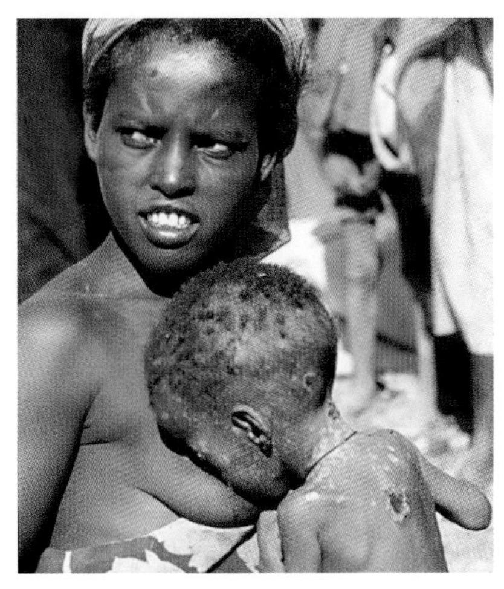

Ntege has kwashiorkor – a disease caused by a lack of proteins in his diet.

Question 4  

When you cut yourself

- some cells are damaged
- some cells die
- other cells are lost when we bleed.

Your body has to make more cells to replace the lost and damaged cells. Cells in your body die all the time and are replaced by new ones.

For example, a red blood cell lasts for only about four months. You make new red blood cells to replace the ones that are worn out.

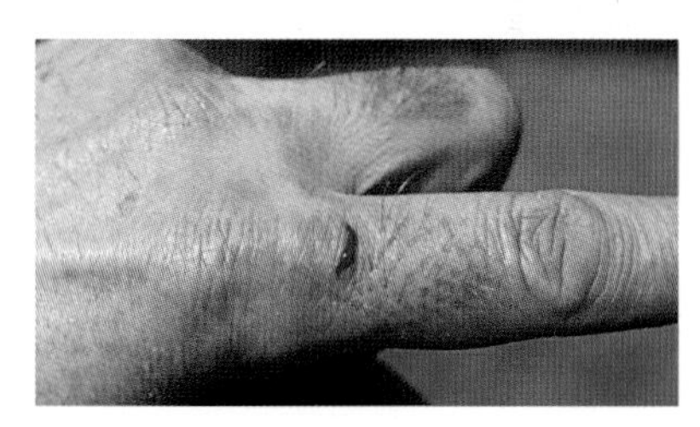

George's finger after an accident.

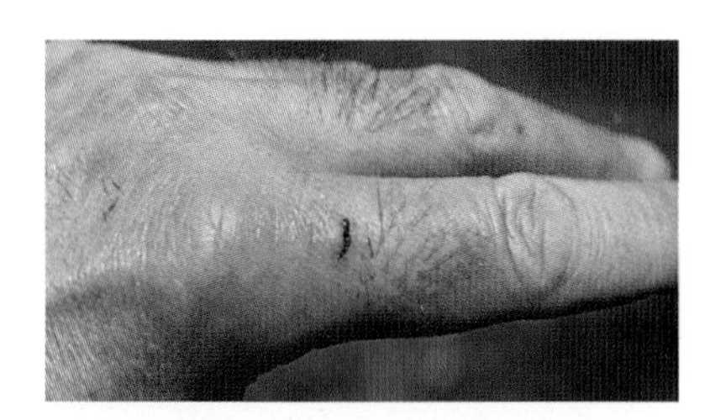

George's finger is healing up.

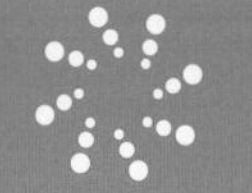

8A.2 Eating for energy, eating for health

You should already know | Outcomes | Keywords

You need **energy** for

- growing
- repairing cells
- moving
- keeping warm.

You release most of this energy from the **carbohydrates** and **fats** in your food.

Carbohydrate foods

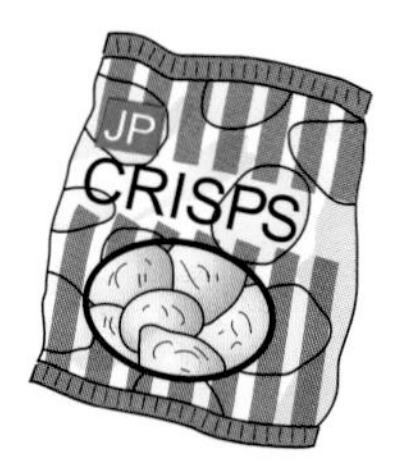

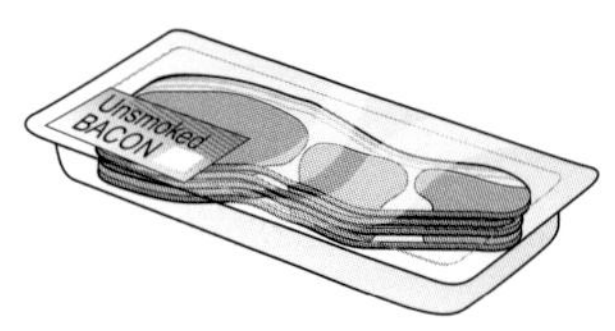

Fatty foods

Question 1 | 2 | 3

Your muscles contract to make you move. Your muscles need energy to do this. So, the more you move around, the more energy you need.

We measure the energy used in kilojoules (kJ). Look at the table.

Activity	kJ per hour
sitting	63
standing	84
walking	750
swimming	1800
walking upstairs	4184
sprinting	5183

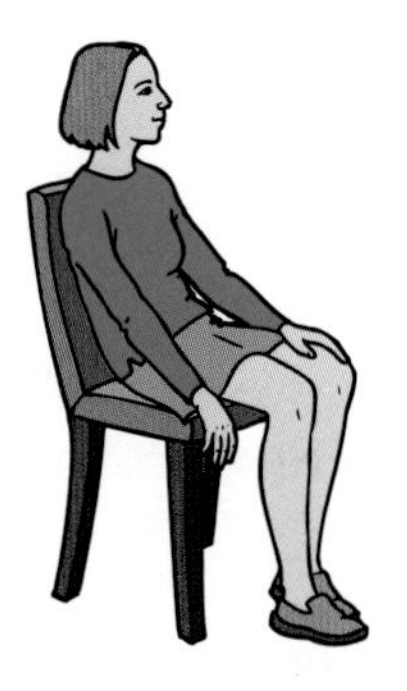

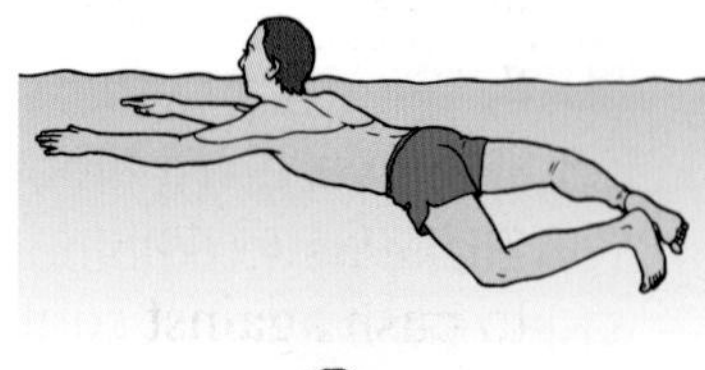

Question 4 | 5

To keep healthy

You need only small amounts of **vitamins** and **minerals**. But they are very important for your health.

In the 1740s, two-thirds of sailors on long voyages died of a disease called scurvy. We now know why.

When sailors were away from land for a long time, they didn't get any fresh fruit and vegetables. So they didn't get any vitamin C. Lack of vitamin C causes scurvy.

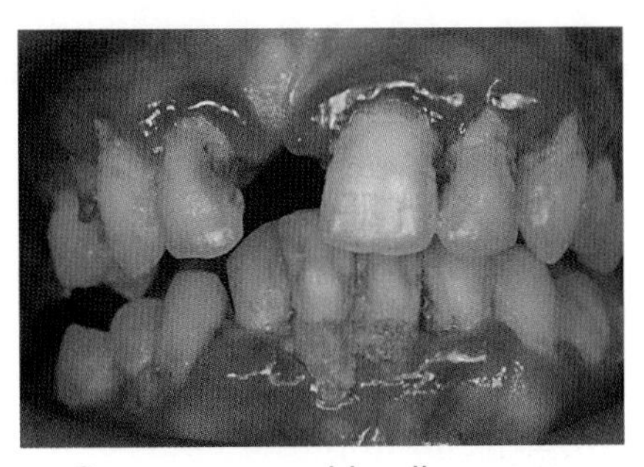

Scurvy causes bleeding gums as well as more serious symptoms.

Minerals are also important.

- Calcium is important for making bones and teeth.
- Iron is important for making red blood cells.

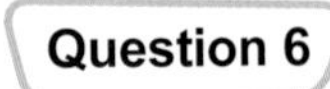 

Question 6 7

The label shows some of the nutritional information from a packet of Sugary Puffs cereal.

NUTRITIONAL INFORMATION	
TYPICAL VALUE per 100 g	
Energy	1620 kJ
Protein	6.5 g
Carbohydrates	86.5 g
(of which sugars)	49.0 g
Fat	1.0 g
Fibre	3.0 g
VITAMINS	
Thiamin (B1)	1.0 mg
Riboflavin (B2)	1.0 mg
MINERALS	
Iron	8.0 mg

Question 8 9 10

Most foods contain a lot of **water**. Think of how watery a melon is!

Even your body is two-thirds water!

Fibre fact file

- **Fibre** is the cellulose in plant cell walls, so fruit and vegetables contain lots of fibre.
- Your body can't break it down, so it goes through your digestive system.
- It gives the muscles of the digestive system something to push against so it helps to keep the food moving.
- Without fibre, you'd be very constipated!

fish

vitamins A, D

egg

vitamins B, D

milk

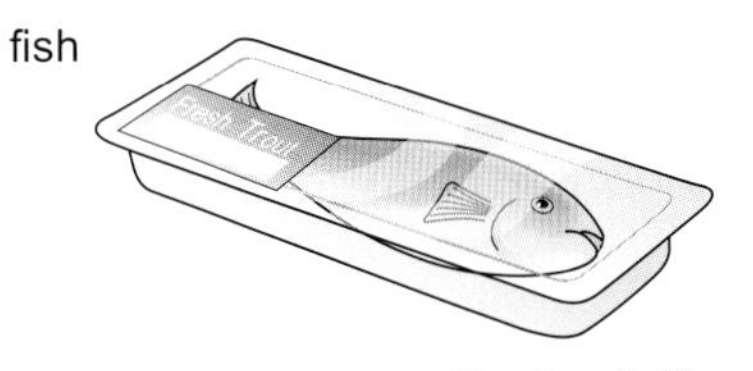

vitamins A, D; calcium

liver

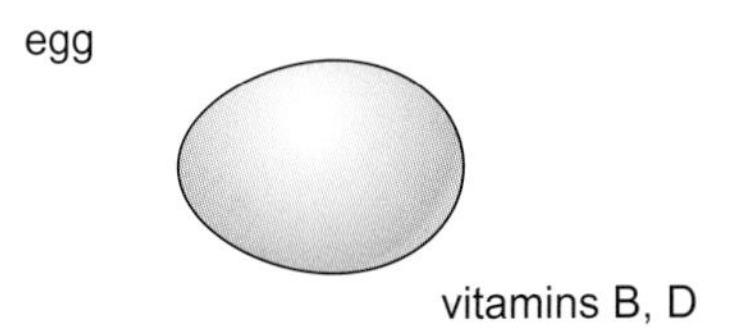

vitamins A, D; iron

vegetables

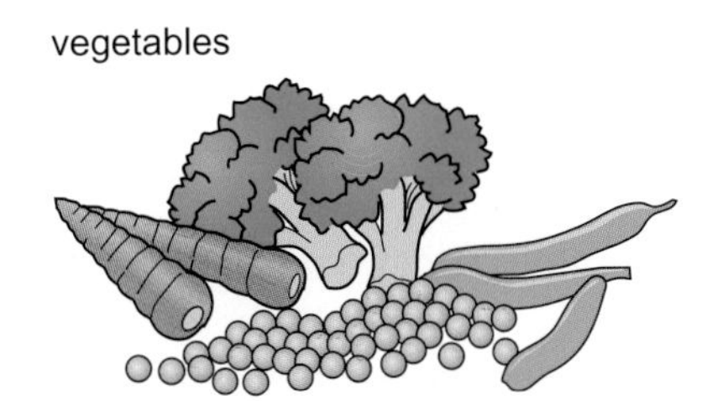

vitamins A, B, C

wholemeal bread

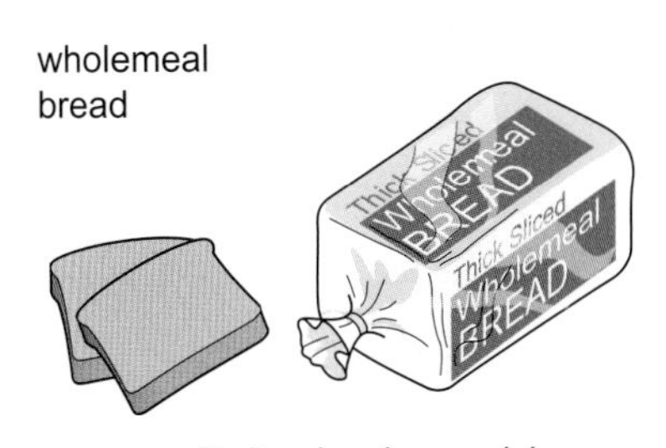

B vitamins, iron, calcium

Foods containing vitamins and minerals.

Question 11

You should already know

Outcomes

Keywords

Water in your diet

Water is an important part of your diet. There is a lot of it in food, more in some foods than others. Having enough water in your body helps to keep it working properly.

You are less likely to get

- headaches
- heatstroke (in hot conditions).

There is also some evidence that having enough water may improve learning.

Drinking water improves test results

This is the claim made following experiments at an Edinburgh primary school. Pupils did better in tests when they could have a drink of water at any time in the school day.

One theory is that water helps nerve impulses to pass. So children with enough water in their bodies learn better.

Question 1 | 2

Finding out how much water is in food

Mrs Tasker asked her class to look up ways of finding how much water there was in various foods. No one found out how to do that exact experiment.

Bryan found one about the amount of water in soil. Using the same idea, he suggested an experiment.

- Find the mass of the food.
- Heat it to get rid of all the water.
- Cool it and find its mass again.
  The loss in mass is equal to the mass of water that was in the food.

Anna found out that they needed to be sure that all the water had gone.

So you heat and cool several times until the mass before heating and the mass after heating are the same. This is called heating to constant mass.

Lisa suggested chopping the food to make it dry faster.

Question 3

Mrs Tasker was pleased with the ideas so far. But she said they needed to find out the best way to heat the food.

Working out the best way to heat the food

Mrs Tasker suggested that they needed to find out which way of heating worked best. So they needed to do some **preliminary tests**.

They tried out their ideas using an apple.

Results of preliminary test.

	Heat over Bunsen flame		Dry on an open shelf at 20 °C		Heat in an oven at 100 °C		Heat in an oven at 300 °C	
Size of apple pieces	cut into eight	chopped up small	cut into eight	chopped up small	cut into eight	chopped up small	cut into eight	chopped up small
Mass at start (g)	140.8	136.4	142.3	143.5	138.6	136.7	133.6	139.1
Mass after 40 mins (g)	12.7	10.1	137.6	135.2	103.5	100.8	10.9	10.1
Mass after 1 day (g)	not done	not done	69.1	67.7	17.3	16.2	9.4	8.4
Mass after 7 days (g)	not done	not done	28.5	28.9	17.3	16.2	not done	not done
Loss in mass (g)	128.1	126.3	113.8	114.6	121.3	120.5	124.2	130.7
% loss in mass	91	92.5	80	80	87.5	88	93	94
Observations	black (burnt)	black (burnt)	brown, mouldy	brown, mouldy	brown	brown	black (burnt)	black (burnt)

The pupils looked at the data. Then they planned their investigation.

Question 4 5 6 7

The pupils decided that the best plan was to

- chop up the food
- find the mass of the food on a digital balance
- heat in an oven at 100 °C
- heat to constant mass.

They also discussed how to make sure that they got reliable results and collected plenty of accurate data.

Question 8 

8A.3 A healthy diet

You should already know | **Outcomes** | **Keywords**

What is a healthy diet?

A balanced, healthy diet contains the correct amount of each food group. We can get a balanced diet in all sorts of ways.

We all need **protein**. We can get it from

- meat – if we like meat and can afford it
- cereals and beans.

A West Indian meal of rice, prawns and vegetables.

Question 1 | **2** | **3**

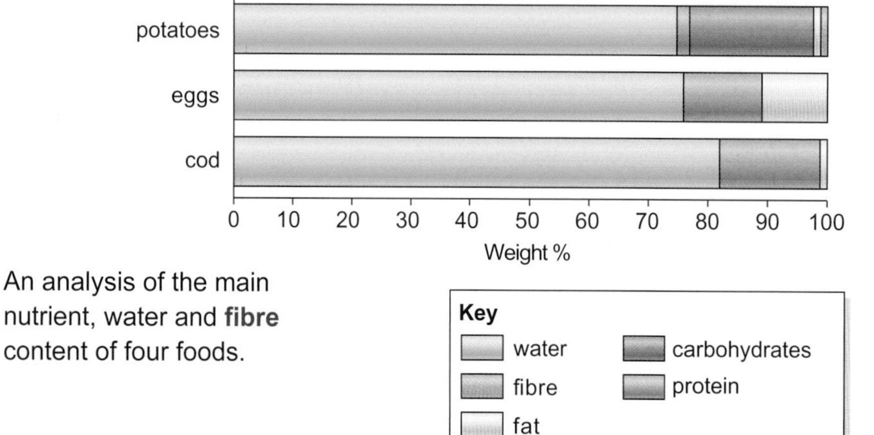

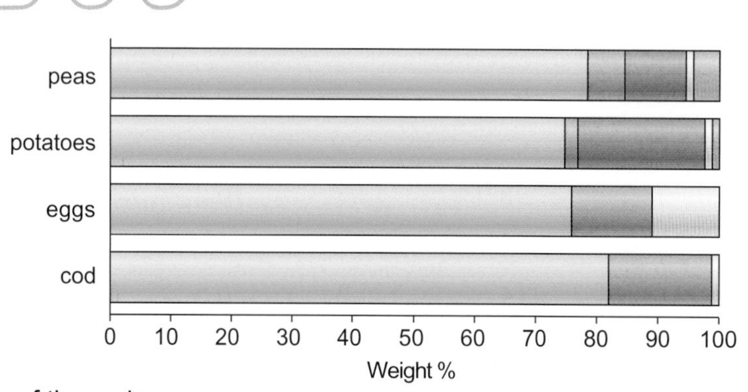

An analysis of the main nutrient, water and **fibre** content of four foods.

A Chinese meal of noodles, prawns and vegetables.

Question 4 | **5** | **6**

A European meal of meat, potato and vegetables.

A healthy, balanced diet is different for different people. The things that affect how much of each food group you need include

- your age
- whether you are male or female
- your body size
- the activities and job you do.

For example, a person who does heavy building work needs more **carbohydrates** and **fats** for energy than a person who sits behind a desk all day.

Question 7 | **8**

Check your progress

8A.4 Modelling absorption (HSW)

You should already know

Outcomes

Keywords

The nutrients from your food pass into your blood.

We say that you **absorb** them.

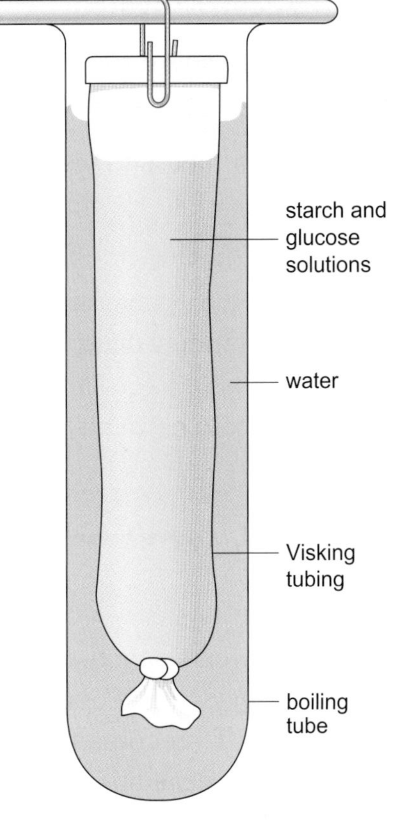

Modelling what happens in your digestive system

In science, we sometimes use **models** to help us understand how things work. Look at the diagram of the Visking tube model.

We can use this model of the gut to find out which substances can pass into the blood and which can't.

Question 1 2 

Now look at the diagram of what happens in your gut. You can see that

- you can absorb: small molecules of vitamins, minerals and some sugars such as glucose
- you cannot absorb: the large, insoluble molecules of fats, proteins and some carbohydrates.

You have to break large molecules down into smaller molecules. We call this process **digestion**. It happens in your digestive system. Then you absorb these small molecules into your blood. Your blood transports them to your cells.

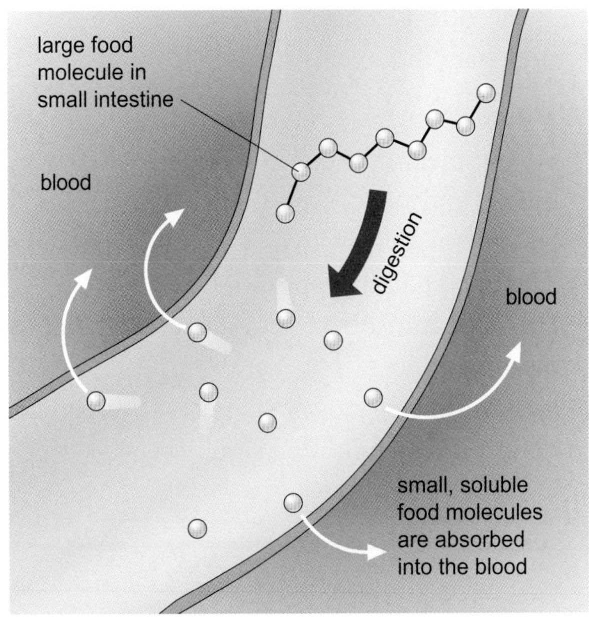

Absorption in the gut.

Question 4 5

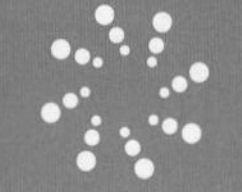

8A.5 Gathering evidence about digestion (HSW)

You should already know

Outcomes

Keywords

Your digestive system

Your food travels through your digestive system

- from your mouth to your anus
- a distance of 8 to 9 metres
- for 24 to 48 hours.

Your digestive system breaks down food into nutrients that you can **absorb**. It has a large surface area to absorb these nutrients.

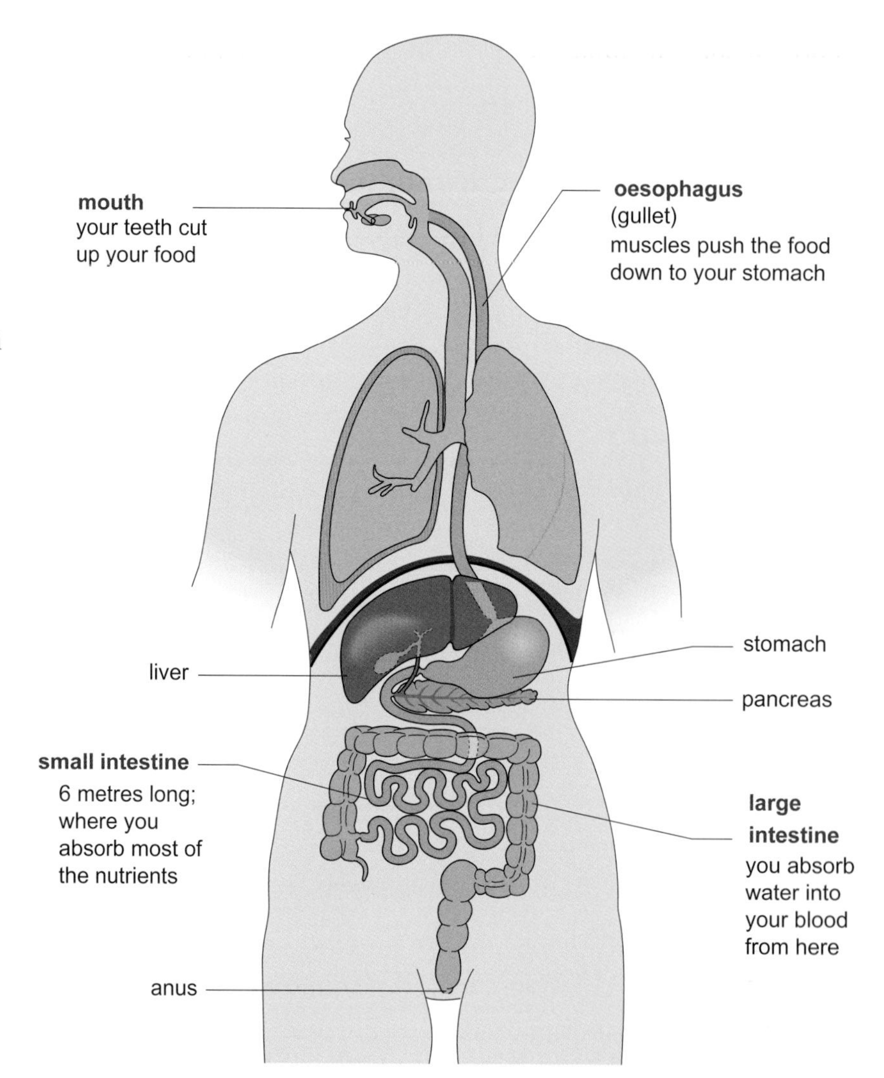

Question 1

Over 3000 years ago, the ancient Greeks found out about the different organs by dissecting dead bodies.

In the 1760s, Lazzaro Spallanzani did experiments on himself.

- He put meat in holes in wooden blocks and swallowed them. He collected the blocks when they passed out of his anus. The meat had disappeared.
- He made himself vomit and showed that the liquid vomit dissolved meat.
- He swallowed food on a piece of thread and pulled the food out before it was fully digested.

Question 2

Lazzaro Spallanzani, an Italian priest.

What enzymes do

We now know that Spallanzani's meat disappeared because **enzymes** had broken it down.

What is produced when different foods break down.

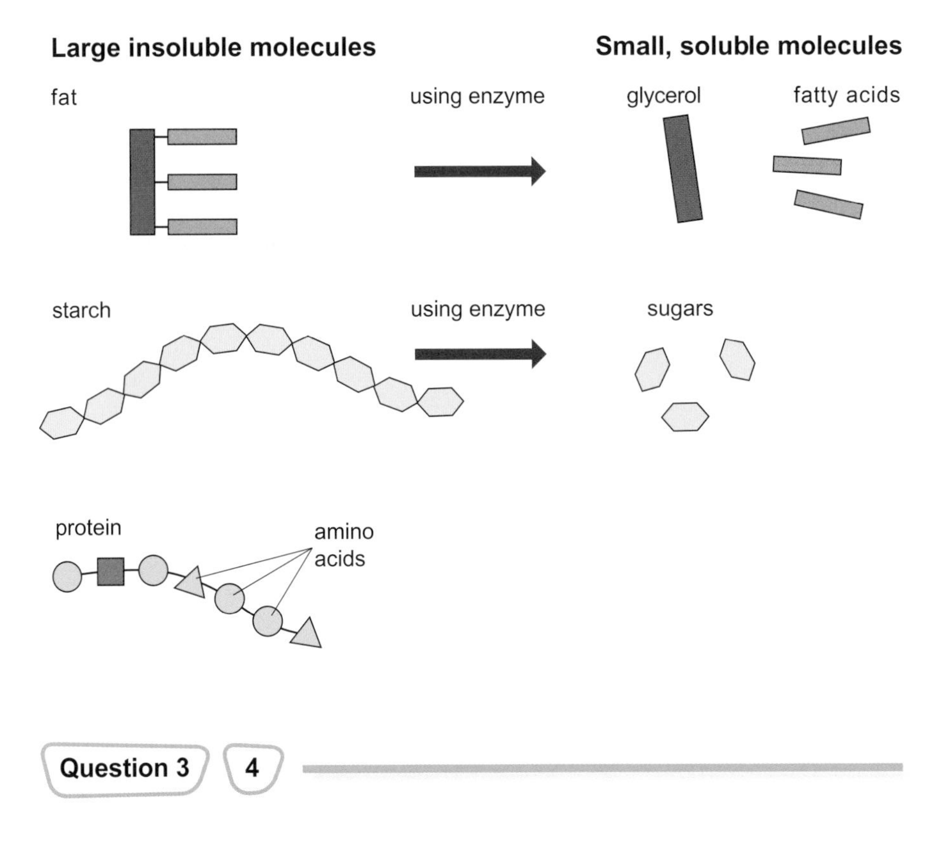

 Question 3  4

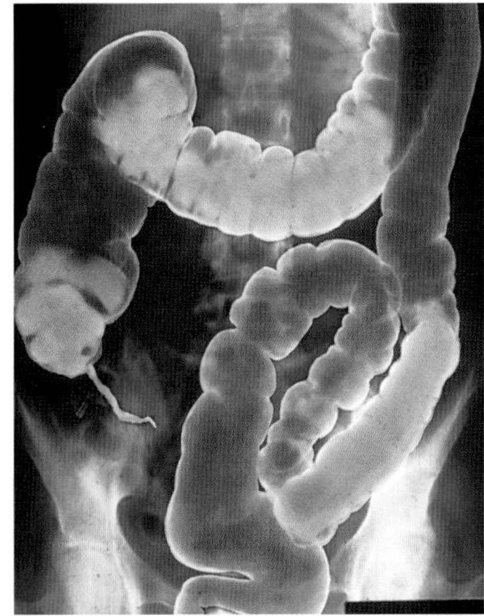

The patient drank a liquid containing a harmless barium compound. This X-ray shows the barium in the large intestine.

Scientists continue to research what happens in the digestive system. They have found out that

- different enzymes break down different foods
- different enzymes work best in different conditions. For example, some work best in acidic conditions, others in alkaline.

Often, doctors and specialist scientists work together to find out more about illnesses. Look at the photographs.

Doctors can use a tiny camera on the tip of an endoscope to photograph inside the digestive system.

Question 5 6 7

8A.6 After digestion (HSW)

You should already know | Outcomes | Keywords

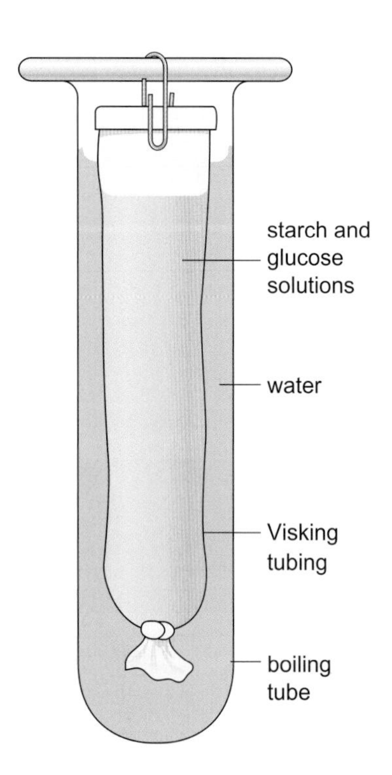

Remember the model gut

The starch molecules are too large to pass through

- the tubing into the water
- your **small intestine** lining.

You break starch molecules down to sugars. Then they can pass through the lining into your bloodstream. Your small intestine is adapted for **absorbing** food.

- It is long.
- It has a thin lining.
- It has a large surface area.
- It has lots of blood capillaries to carry away nutrients.

Question 1 2 3 4

Every cell of your body needs nutrients. Your cells need them

- for growth
- for repair
- as an energy source.

So your bloodstream carries nutrients to all parts of your body. It carries the nutrients in solution in the blood plasma – the liquid part of blood.

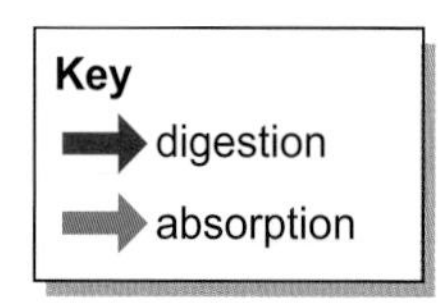

Question 5 6

What happens to the nutrients

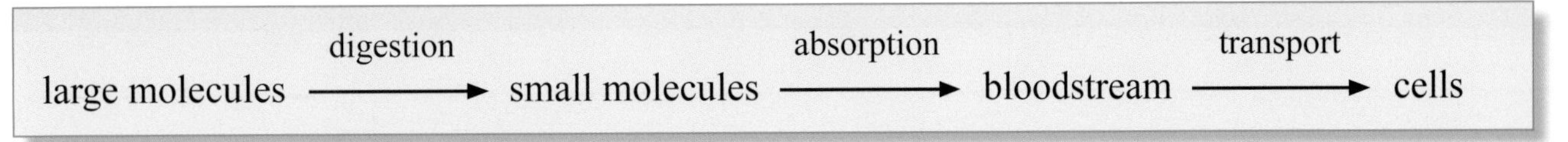

What happens to the products of digestion.

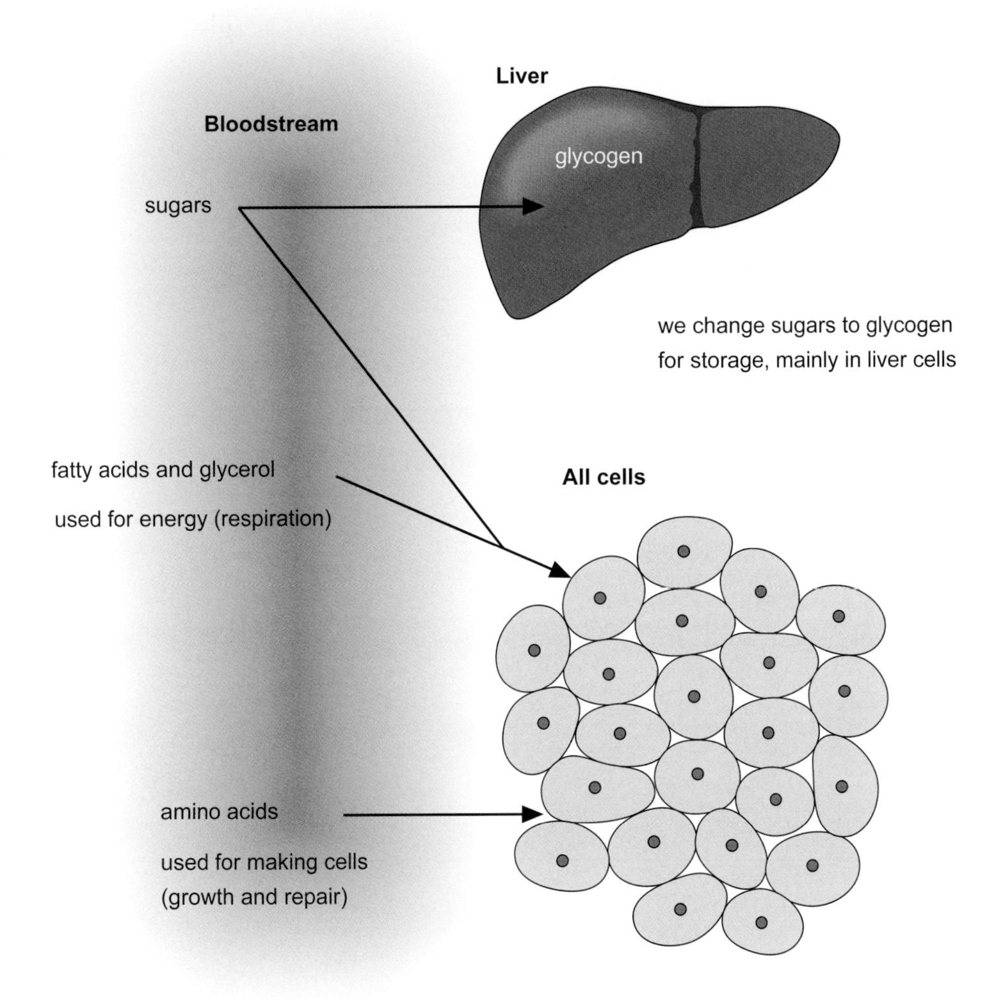

Question 7 **8**

What happens to the waste

All the undigested food, including **fibre**, is got rid of in **faeces**.

We say that we **egest** it.

Faeces are mainly fibre, water and bacteria.

Review your work

Summary ➡

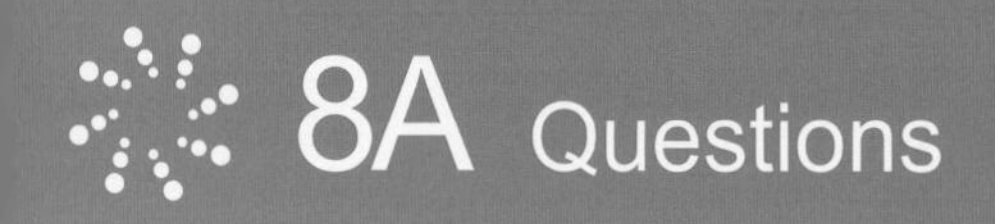

8A Questions

8A.1

1 Write down:
   a two reasons you need to make new cells;
   b three uses for energy in your body.
2 Find out why you need
   a one named vitamin
   b one named mineral.
3 On average, a pregnant woman needs 76 g of protein per day. Other women need less.
   a Explain why a pregnant woman needs extra protein.
   b Find out how much protein a woman normally needs.
4 Write down two differences between the children in the photographs.
5 Write down two foods that will improve Ntege's health if he can get them.
6 Some children don't eat enough protein foods. Suggest some long-term effects of this.
7 Look at the pictures of George's wound. It slowly healed up. Describe how.
8 Wounds take longer to heal if the person doesn't eat enough protein foods. Suggest why.

8A.2

1 Write down four energy foods that you eat.
2 Children grow more slowly than they should when they don't get enough energy foods. Why is this?
3 Nathan is 24 years old. He is no longer growing. What will happen to him if he doesn't get enough energy foods?
4 Our bodies still use energy when we are sitting still. What do they use the energy for?
5 We need more energy when standing up than sitting down. Suggest why.

continued

6 Look at the photograph. Describe some effects of scurvy.

7 Write down

- **a** one food that contains vitamin C
- **b** one food that contains vitamins A and D, and the mineral iron
- **c** one food that contains calcium.

8 What is the main nutrient in Sugary Puffs?

9 Which mineral is found in Sugary Puffs?

10 Write down one vitamin in Sugary Puffs.

11 Explain why you need water and fibre in your diet.

8A.HSW

1 Look at the newspaper article about drinking water.
What evidence is there of a link between drinking enough water and learning?

2 In your group, discuss what you would do to investigate the effects on learning of drinking enough water.

3 Look at the speech bubbles. All three ideas were a useful part of the experiment plan.
Explain why each idea was useful.

4 What two kinds of preliminary work did the class do before they planned their investigation?

5 The pupils rejected heating over a Bunsen burner and in an oven at 300 °C because the apple lost more than just water.

- **a** What evidence is there that more than just water was lost?
- **b** Suggest what else was lost.

6 Suggest two problems of drying the apple at 20 °C.

7 These tests didn't show whether chopping up the apple made a difference to the time taken to dry the apple.
What extra tests can the class do to find out the answer?

8 Look at the pupils' plan.
Suggest reasons for each step.

9 In your group, discuss other ideas that the pupils probably used to make their investigation

- **a** safe
- **b** reliable.

8A.3

1 Rice provides most of the energy in the West Indian meal. What provides most of the protein?

2 Prawns provide most of the protein in the Chinese dish. What provides most of the energy?

3 Which part of the European meal contains most of the proteins and fat?

4 Which of the foods on the graph contains the most
 a water?
 b protein?

5 What is the main nutrient in potatoes?

6 Which nutrients in the graph are not in cod?

7 Suggest what the following people need to eat.
 a Mmapula, a 13-year-old girl living in South Africa.
 b Steve, a professional footballer in the UK.

8 Janet is breastfeeding her baby. Find out what she should eat and drink.

8A.4

1 Which part of the model represents the blood?

2 What does the Visking tubing represent?

3 Later, there is glucose in the water around the Visking tubing but no starch. Explain why.

4 a Write down two molecules that can pass into your blood.
 b Write down two molecules that cannot pass into your blood.

5 Write down three substances that you can absorb without digesting them.

8A.5

1 List, in order, the parts of your digestive system that your food travels through.

2 a What happened to the meat in the wooden blocks that Spallanzani swallowed?
 b Where did this happen?

continued

3 Choose the correct conclusion (A–D) for Spallanzani's experiments.
   A Meat breaks down to amino acids in the stomach.
   B The juices in the stomach break down meat.
   C Proteins break down into amino acids that can be absorbed.
   D Meat is broken down entirely in the stomach.

4 Proteins break down into amino acids.
   Draw a diagram to show what the protein in the diagram looks like when it is broken down.

5 Write down two things that scientists have found out about enzymes.

6 **a** Find out the connection between an endoscope, fibre optics and a stomach ulcer.
   b Find out one reason why doctors take X-rays of a patient's digestive system.
   In your group, agree on a few sentences about each of **a** and **b**.

7 To find out what is wrong with a patient, doctors use tests and tools. These are developed and used by various specialist scientists working together with doctors.
   a Find out what specialisms were involved in the development of either X-rays or endoscopes. Combine the group's ideas into a list.
   b Sharing ideas and collaborating are important in science. In your group, discuss the reasons.

8A.6

1 What can you put in the Visking tubing to break down the starch into sugar?

2 What happens to the sugar?

3 Your saliva breaks down starch.
   What does this tell you about your saliva?

4 In your group, discuss ways in which the model is different from a real small intestine.
   Write down your answers.

5 Write down two nutrients that can pass into your blood.

6 Write down two groups of nutrients that you don't need to digest.

7 Describe two things that can happen to sugars after they pass into the blood.

8 Write down one kind of cell that uses lots of sugars.
   Explain your answer.

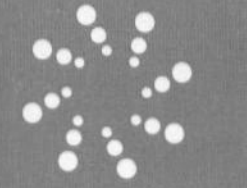

8B.1 How cells use food

Scientific enquiry

You should already know

Outcomes

Keywords

Your cells use food as a source of materials to grow, and for energy.

You use energy to move, grow and keep warm.

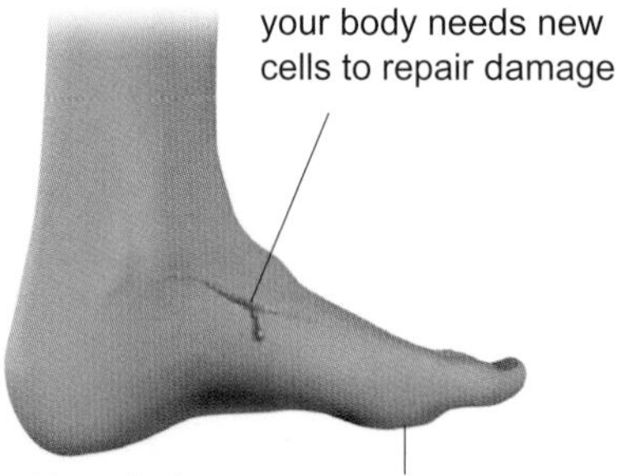

You need glucose for energy and amino acids to make proteins for new cells.

Gail's muscle cells use up more glucose to release extra energy when she runs.

This is a high-energy drink for sports players. It contains a lot of glucose.

Question 1 2 3 4

Releasing energy

Glucose supplies your cells with **energy**. So it is your body's fuel. But your cells don't burn glucose. They break it down to release the energy a bit at a time.

We call this **respiration**.

Because your cells normally use **oxygen** from the air when they respire, we call this **aerobic** respiration.

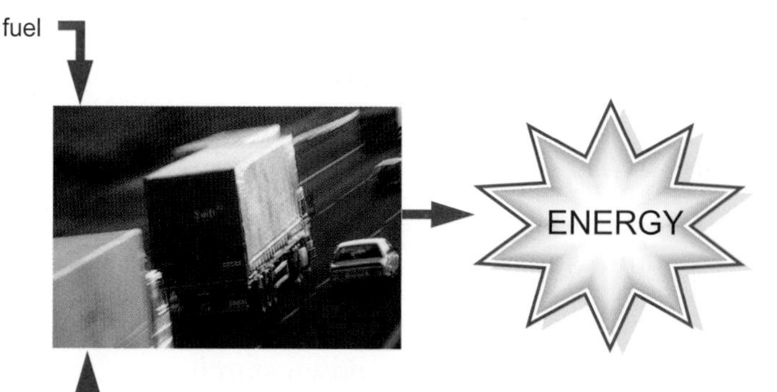

In cars and lorries, fuel is burnt in oxygen to release energy. Your cells also use oxygen to release energy from their fuel.

The word equation for aerobic respiration is:

glucose + oxygen → **carbon dioxide** + water + energy

This equation doesn't show that the glucose breaks down a bit at a time. It just shows the reactants and the products.

Question 5 6

How oxygen reaches your tissues

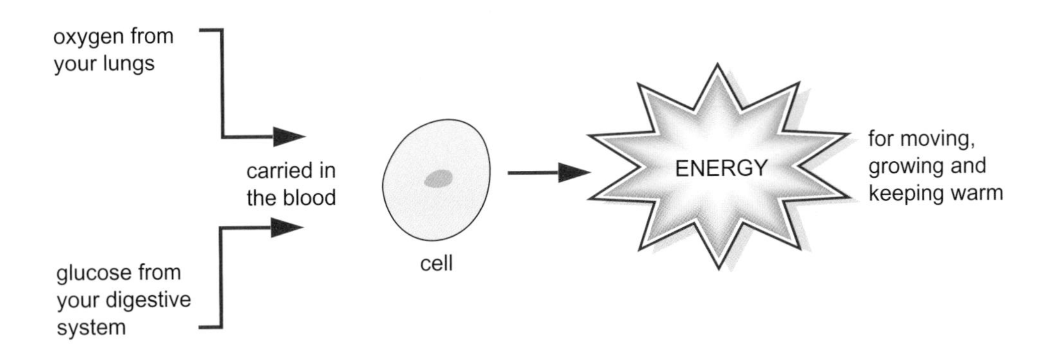

Air contains oxygen. You take air in and out of your lungs. This is called **breathing**. Some of the oxygen from the air in your lungs passes into your blood.

Your blood then carries the oxygen to your tissues. It passes

- from blood into **tissue fluid** (a liquid that surrounds all cells)
- from tissue fluid into cells.

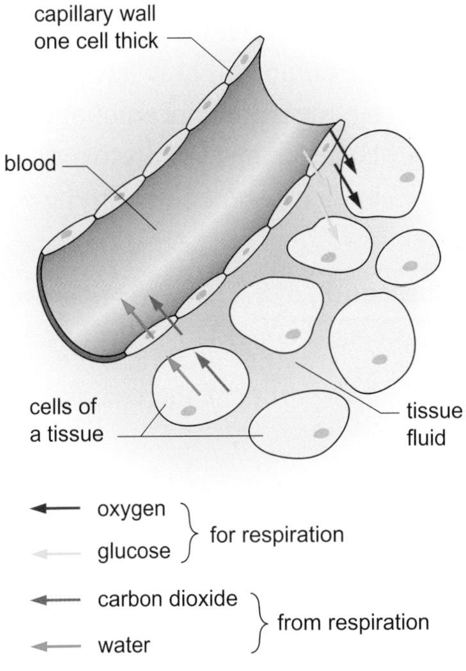

Exchange of materials between cells and the blood.

More about exchanges

Dissolved substances and gases **diffuse** from where they are in high concentrations to where they are in low concentrations. We call the process diffusion.

This is how substances pass in and out of your blood and your cells all the time.

	In your tissues	In your lungs
These go into your blood	carbon dioxide	oxygen
These go out of your blood	oxygen	carbon dioxide

When you breathe out, you get rid of the carbon dioxide that passes out of the blood in your lungs.

Question 10

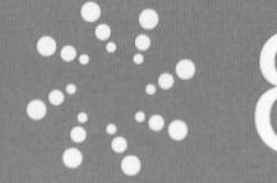

8B.2 How ideas about blood circulation changed (HSW)

You should already know

Outcomes

Keywords

The story of blood circulation

Ideas changed.

- In the 2nd century BC, the Chinese knew that blood circulated.
- The ancient Greeks and Romans thought that blood moved back and forth like the tides in the sea.
- In the 13th century AD, an Arab doctor called Ibn al-Nafis worked out how blood circulates through the **heart** and lungs.
- William Harvey gathered lots of **evidence** about circulation.
  - He observed the hearts of many different mammals.
  - He compared how hearts and pumps work.
  - He measured the amount of blood leaving the heart.
  - He predicted the discovery of tiny vessels between **arteries** and **veins**. But he couldn't see them.

William Harvey is often credited with the discovery of the double circulation of blood through the heart.

Question 1 | 2

Harvey's experiment showed that blood travels in one direction in veins.

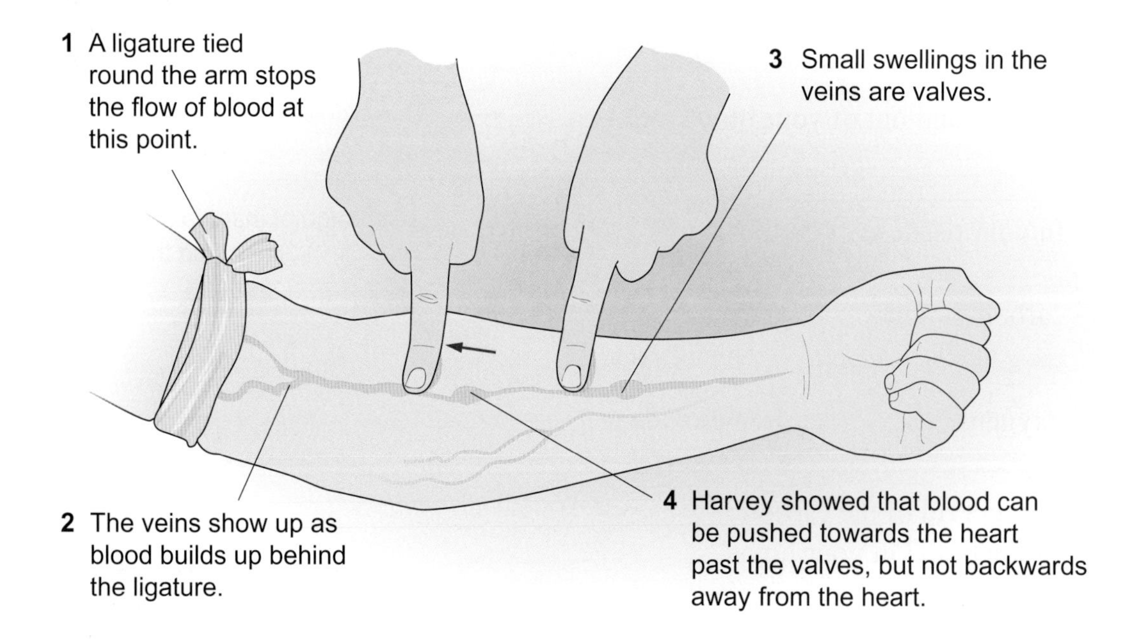

Finding the link between arteries and veins

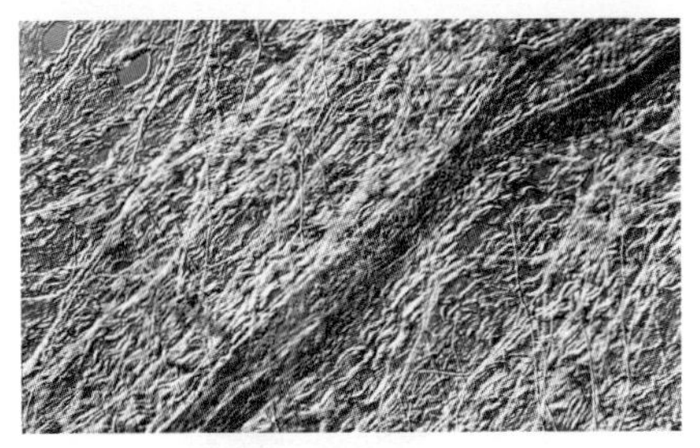

Using a microscope, an Italian scientist called Marcello Malpighi discovered the link some years later. He found the tiny blood vessels that we call **capillaries**.

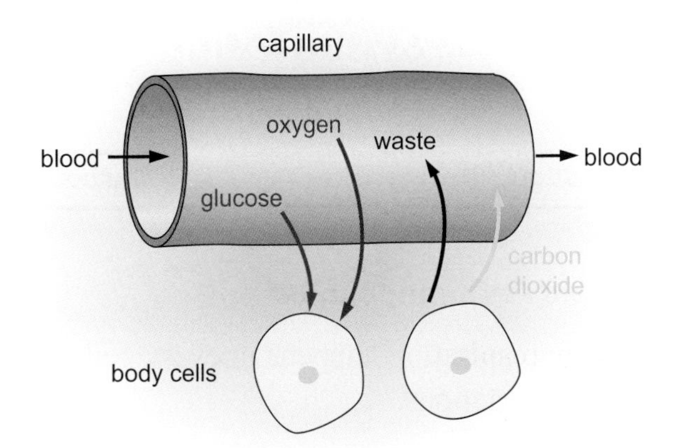

Question 3 **4** **5** **6**

Now we know a lot more

Malpighi knew that blood left the heart in arteries, that arteries split up into tiny tubes called capillaries in your tissues and that capillaries join up to form veins that take blood back to your heart.

We now know that capillaries are where substances go in and out of your blood.

Question 7 **8**

Your heart

Your heart is a muscular pump. Heart muscles squeeze blood to move it around your body.

The two sides of your heart pump blood out at the same time. So we say that it is a double pump.

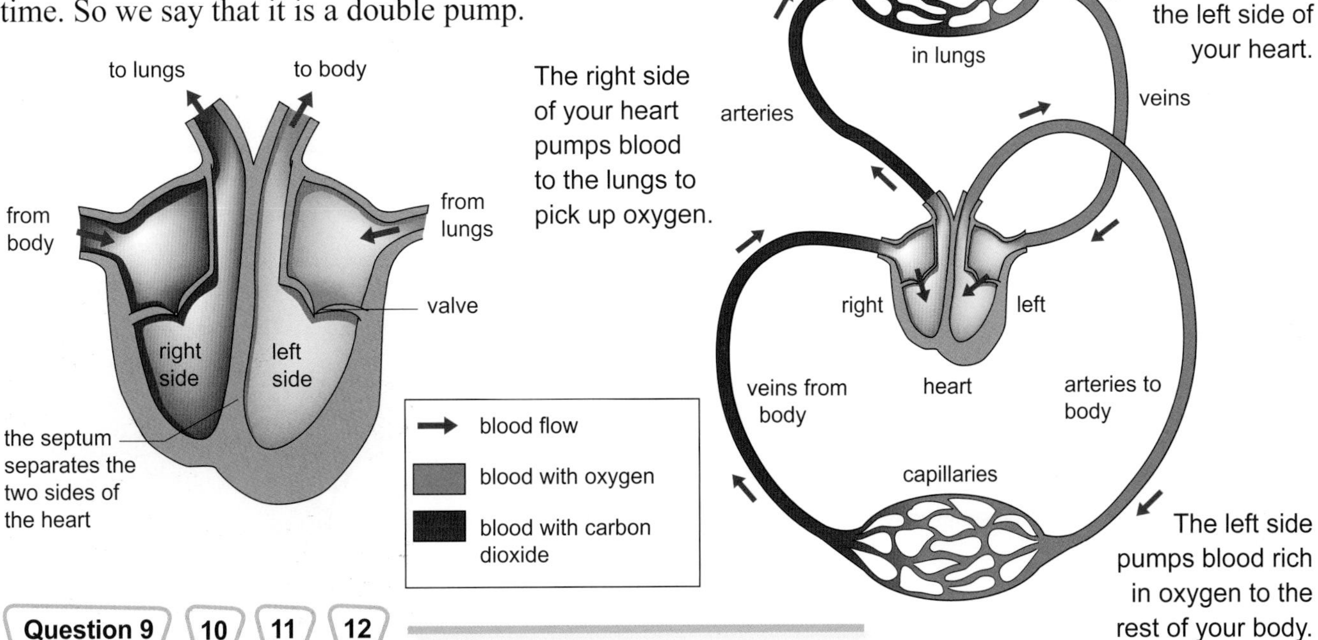

Question 9 **10** **11** **12**

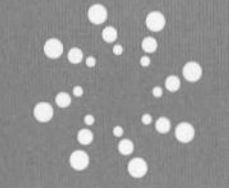

8B.3 Oxygen supplies to cells (HSW)

You should already know

Outcomes

Keywords

Aerobic respiration in cells

Aerobic respiration happens in every cell. It is a series of chemical reactions.

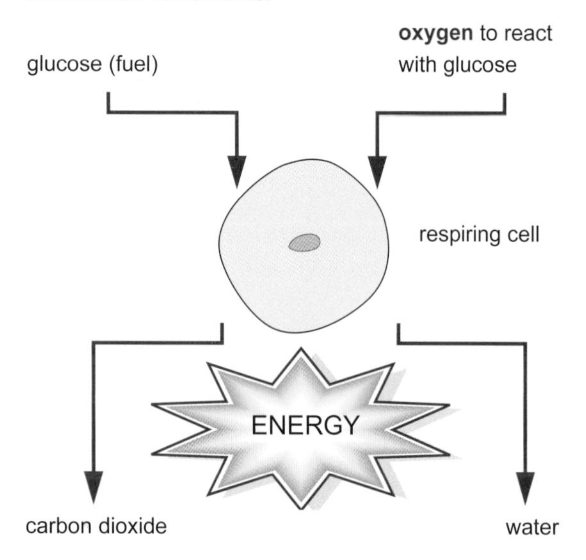

An aerobics exercise class.

Question 1 | 2

Sometimes there isn't enough oxygen in the air

This climber is 4500 m above sea level. At high altitudes like this, the air molecules are more spread out.

The climber is taking in less oxygen than normal with each breath. So his cells are getting less oxygen than they need.

Lack of oxygen leads to tiredness, a bad headache and difficulty in concentrating.

Question 3 | 4

The air in a passenger aeroplane is kept pressurised.

So the amount of oxygen in the air is similar to that near the ground.

Question 5

Sometimes lung damage reduces oxygen uptake

Lung tissue is thin and delicate. So it is easily damaged. Damage makes the surface area of the **lungs** smaller. So less oxygen goes into the blood.

Jack's story

I spent 25 years working in a coal mine. As a result of breathing in dust for many years, I developed a bad cough, then miner's lung (pneumoconiosis). The stretchy tissue inside my lungs was damaged. It was being replaced with fibrous tissue that didn't let my lungs stretch as I breathed in.

I was finding it harder and harder to breathe and had to give up work. Now I have to use oxygen from a cylinder. I am so short of breath that I can only walk about 100 metres. We didn't have dust masks in my day.

Question 6 7

Smoking damages the lungs too

Smokers often cough because tobacco smoke irritates their air passages. Coughing damages the lungs. People with damaged lungs get out of breath easily. They can't take in enough oxygen.

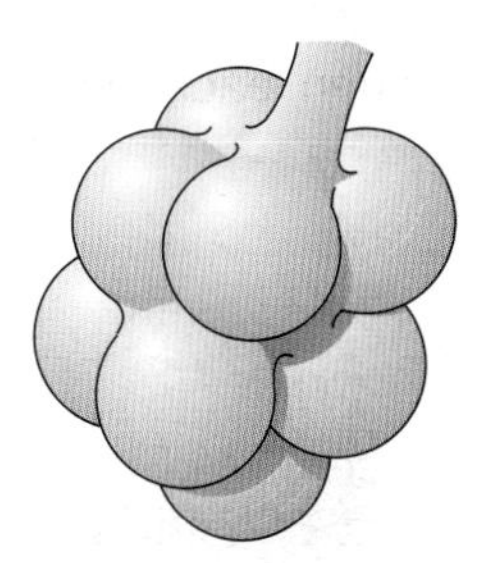

Healthy lung tissue.

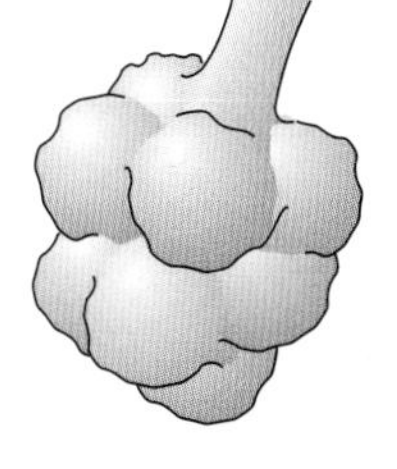

Damaged lung tissue.

Question 8 9

Check your progress

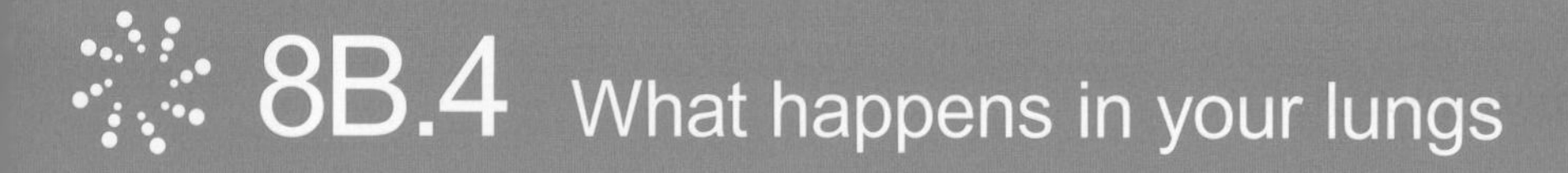

8B.4 What happens in your lungs

You should already know

Outcomes

Keywords

Obtaining oxygen, getting rid of carbon dioxide

You should already know why your lungs are so important and why they are so easily damaged.

Remember that you get the oxygen you need from the air.

You **breathe** air in and out of your **lungs**:

- oxygen from the air diffuses into your blood
- waste carbon dioxide passes from your blood into the air.

We call this **gas exchange**.

So there is less oxygen and more carbon dioxide in the air that you breathe out than in the air that you breathe in.

Question 1 | 2

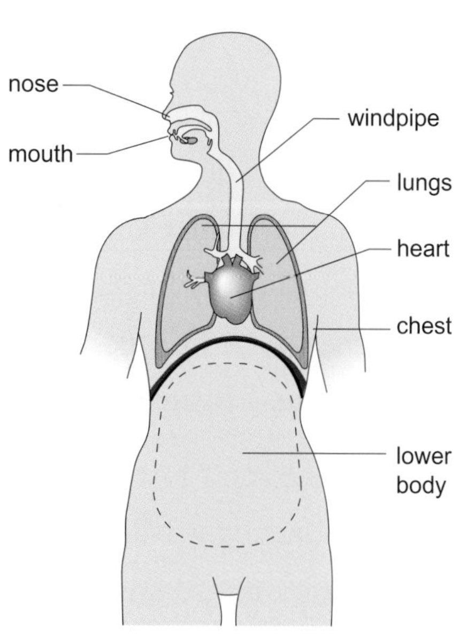

The air passages to the lungs.

How gas exchange happens

Inside the lungs are millions of tiny air sacs called **alveoli**.

Alveoli make the lungs feel spongy. Alveoli have features that make them suitable for gas exchange. They have

- a very large surface area
- walls that are only one cell thick
- lots of capillaries with walls only one cell thick
- a moist lining that gases can dissolve in.

These features allow dissolved gases to pass quickly between the air in the alveoli and the blood in the capillaries.

The blood carries oxygen to the body cells and brings waste carbon dioxide from the body cells to the alveoli.

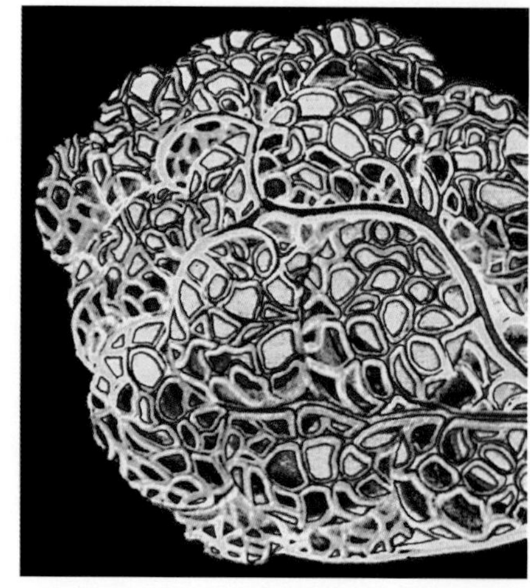

Alveoli in the lungs.

Question 3  4  5

So the stages in getting oxygen to your cells and getting rid of carbon dioxide are as follows.

- Breathing: you breathe air in and out of your lungs.
- Gas exchange: gases are exchanged between the air in the alveoli and the blood.
- Transport: transport of gases in the bloodstream between the lungs and the cells.

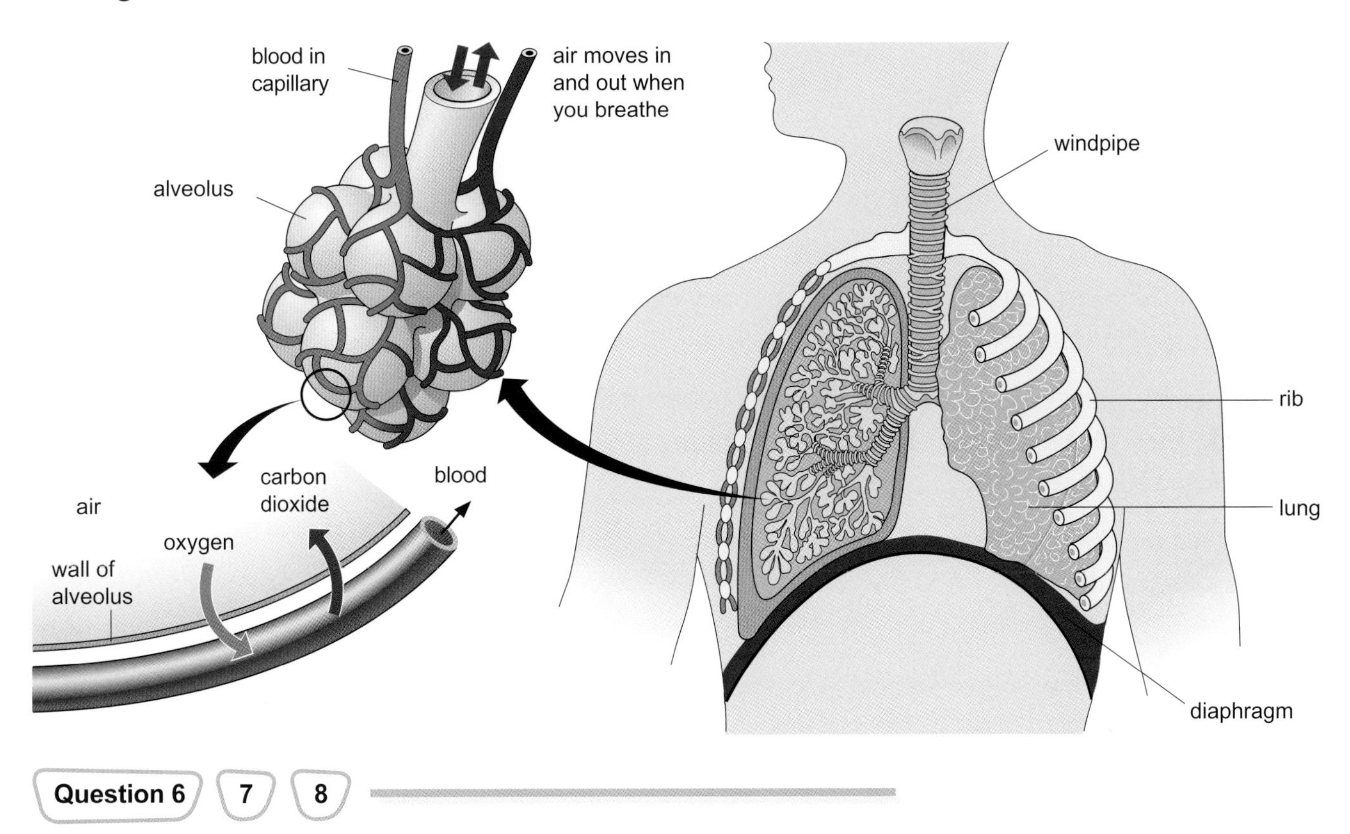

Question 6 7 8

Smoking affects …

… gas exchange

You already know that coughing damages alveoli and reduces the surface area of the lungs. So, less oxygen passes into the blood.

… and transport of oxygen

Red blood cells normally carry oxygen. But they carry carbon monoxide even more easily. Tobacco smoke contains carbon monoxide.

So, on average, smokers carry less oxygen in their blood than non-smokers do.

Question 9 10

8B.HSW Collaborating to combat lung disease

You should already know | Outcomes | Keywords

Investigating diseases

Doctors and public health officials work together to find out the reasons for diseases. When people work together, we say that they **collaborate**.

We call the study of the **causes** of diseases in populations **epidemiology**. When doctors know the causes of a disease, they can work to prevent it.

Diseases such as heart and lung disease have many causes. Researchers gather data about medical histories and lifestyle.

Then they analyse the data. They look for links between the disease and lifestyle factors such as diet and smoking. We call these **risk factors**.

To prove that a risk factor causes a disease, they need to find out how it causes that disease.

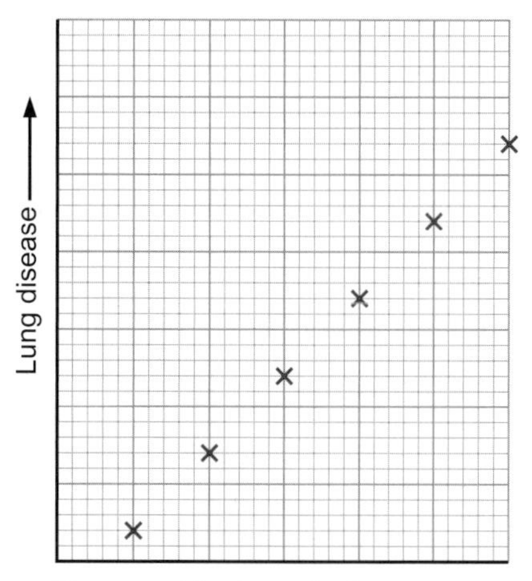

This graph shows the kind of link that researchers look for. It shows that mining might be a cause of lung disease.

Question 1 | 2

Risk factors for lung disease

- Smoking tobacco, including passive smoking.
- Breathing in dust and chemicals at work.
- Indoor air pollution, e.g. from open fires.
- Outdoor air pollution.
- Frequent respiratory infections in childhood.

Worldwide collaboration

The World Health Organisation (WHO) works on worldwide health issues.

It has a group (called GARD) working on lung disease. Members include institutions, doctors, health educators, scientists and statisticians from around the world.

- They are gathering and analysing data
- They are developing standard research methods – they can combine data if they collect the same kind of data in different countries
- They are providing evidence for health improvement policies and helping to carry them out.

By working together, they hope to change things more quickly.

The WHO is part of the United Nations.

Worldwide data

Doctors now group lung diseases together as

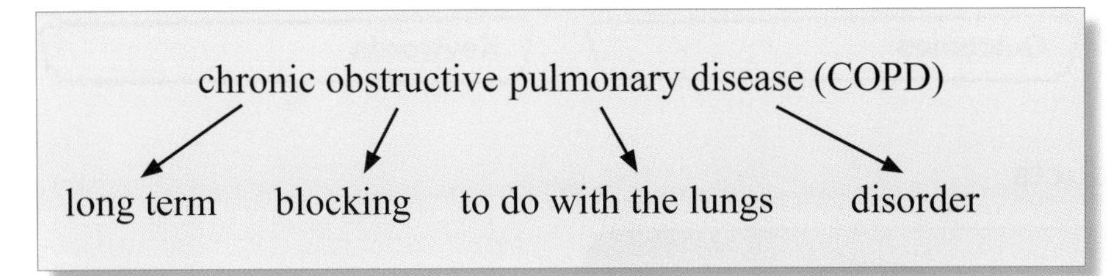

The data brought together by the WHO shows that, worldwide

- over 3 million people die from lung diseases every year – about 5% of all deaths
- over 90% of deaths from COPD are in poorer countries
- 80 million people are known to have moderate to severe COPD (600 000 are in the UK)
- millions more have COPD but don't know
- COPD affects men and women equally.

Question 3 4 

Members of GARD are working to reduce the risk factors that people are exposed to, such as

- tobacco smoke
- factory and mine dust
- particulates such as soot.

Reducing smoking and air pollution will reduce lung disease. This will help families, communities and the economy.

However, it will take many years for this work to have an effect.

If we do nothing, the WHO suggests that

- deaths from COPD will increase by more than 30% in the next 10 years
- absence from work because of COPD will increase.

Question 6  

Factfile

COPD includes

- chronic bronchitis
- emphysema
- asbestosis
- pneumoconiosis
- tumours

Signs and symptoms

- shortness of breath
- coughing and phlegm
- bluish skin (lack of oxygen)

Open fires for cooking cause indoor air pollution.

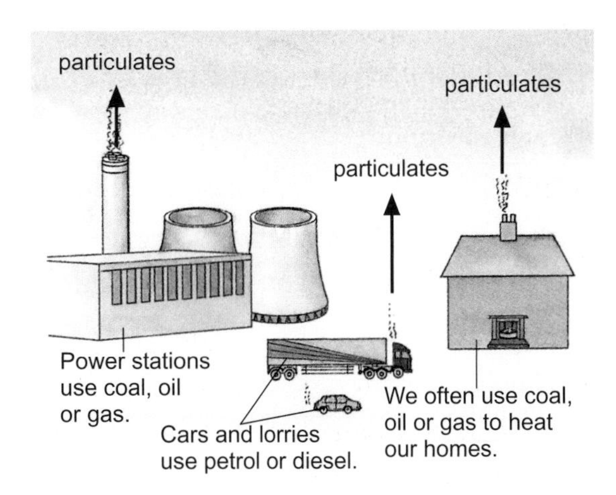

Particulates such as soot from vehicles and industry cause outdoor air pollution that can affect our lungs.

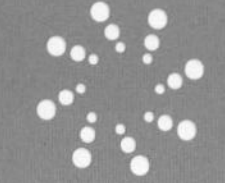

8B.5 Comparing inhaled and exhaled air

You should already know

Outcomes

Keywords

Respiration makes waste products

Question 1 2 3

All cells make waste products when they respire.

The waste products of aerobic respiration are

- **carbon dioxide**
- **water**.

Carbon dioxide is poisonous so you get rid of it in the air you exhale (breathe out).

Question 4

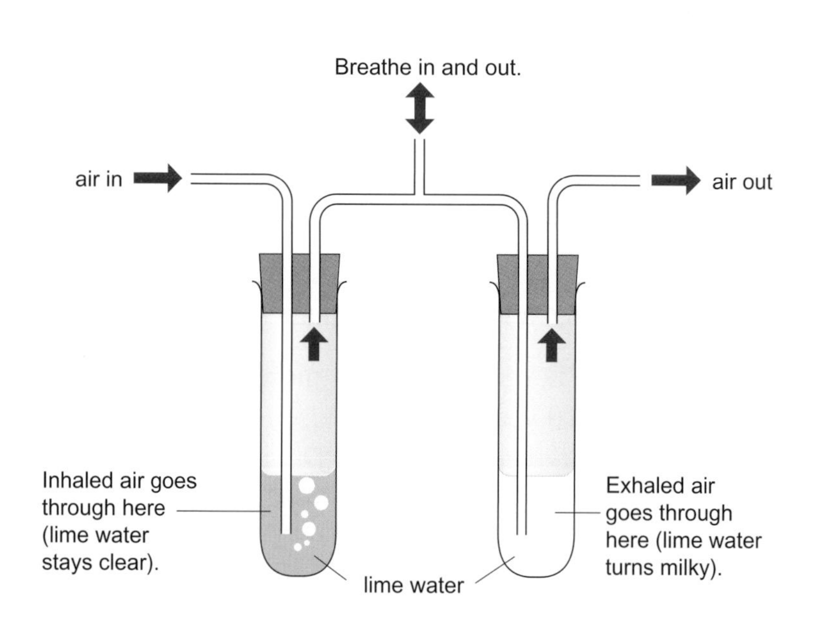

Comparing the amounts of carbon dioxide in inhaled and exhaled air.

Peter did an experiment to compare the amounts of gases in air breathed in and out. Look at his results.

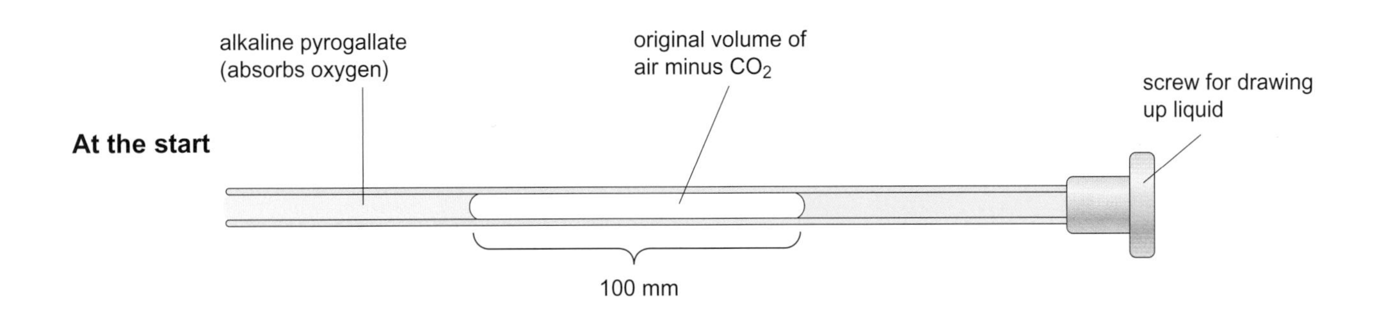

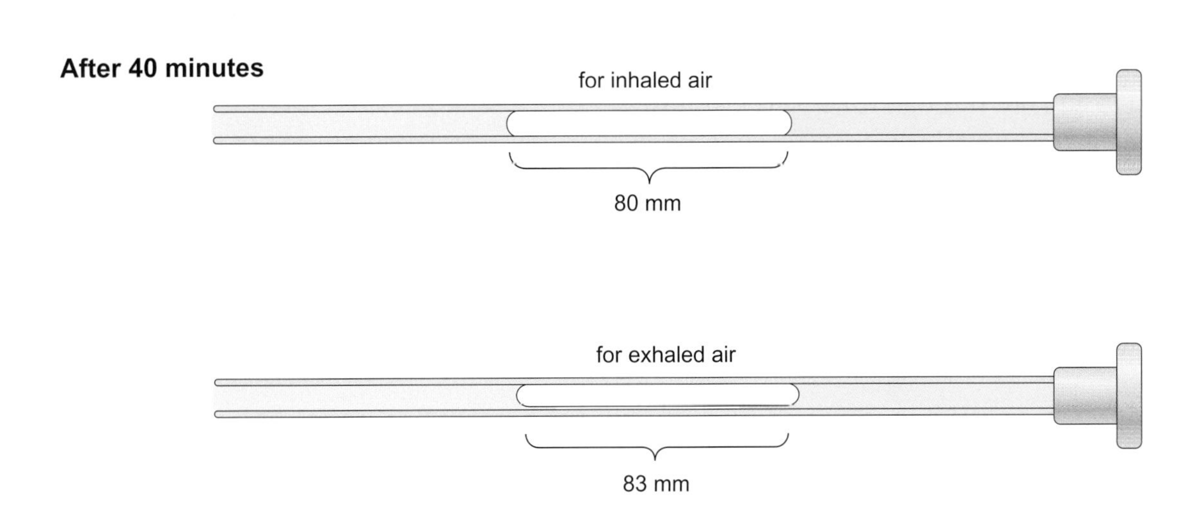

Question 5

Gas	Air breathed in (%)	Air breathed out (%)
oxygen	20	17
carbon dioxide	0.03	4
nitrogen	79	79
water vapour	varies	saturated

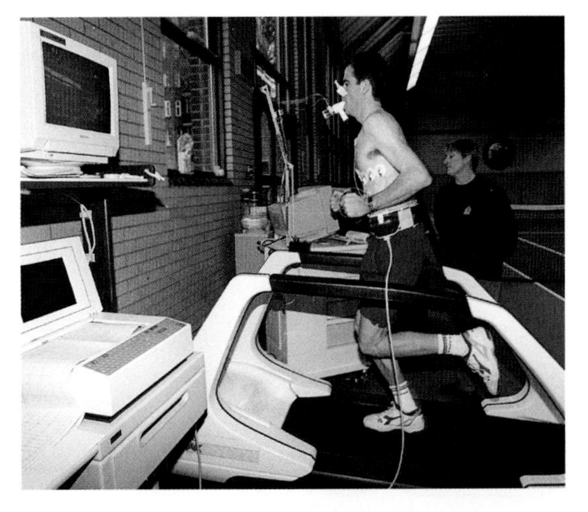

This athlete is measuring the gases in his exhaled air.

Question 6   9

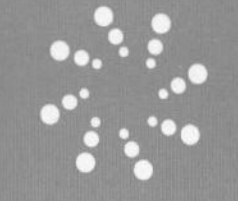

8B.6 Other living things respire too (HSW)

You should already know

Outcomes

Keywords

Finding evidence for respiration

The cells of living things

- release **energy** to carry out their life processes
- usually use oxygen and produce **carbon dioxide**.

Carbon dioxide production is a good way of finding out if **respiration** is happening.

When you breathe in and out through lime water, you find that the air you breathe out contains more carbon dioxide than the air you breathe in.

But you can't ask a seed or a woodlouse to breathe in and out through lime water!

The diagrams show what you can do.

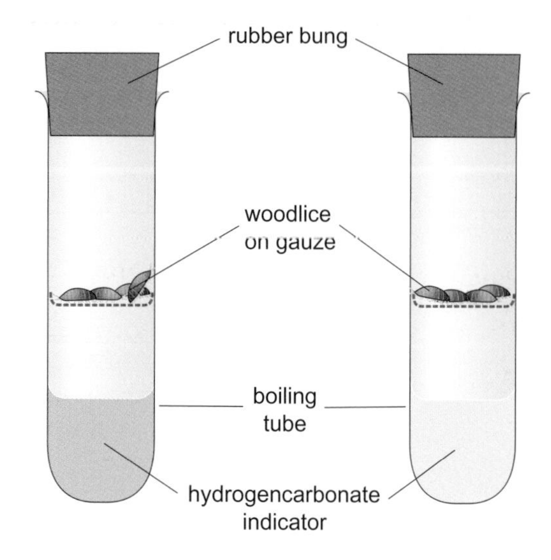

Experiment 1

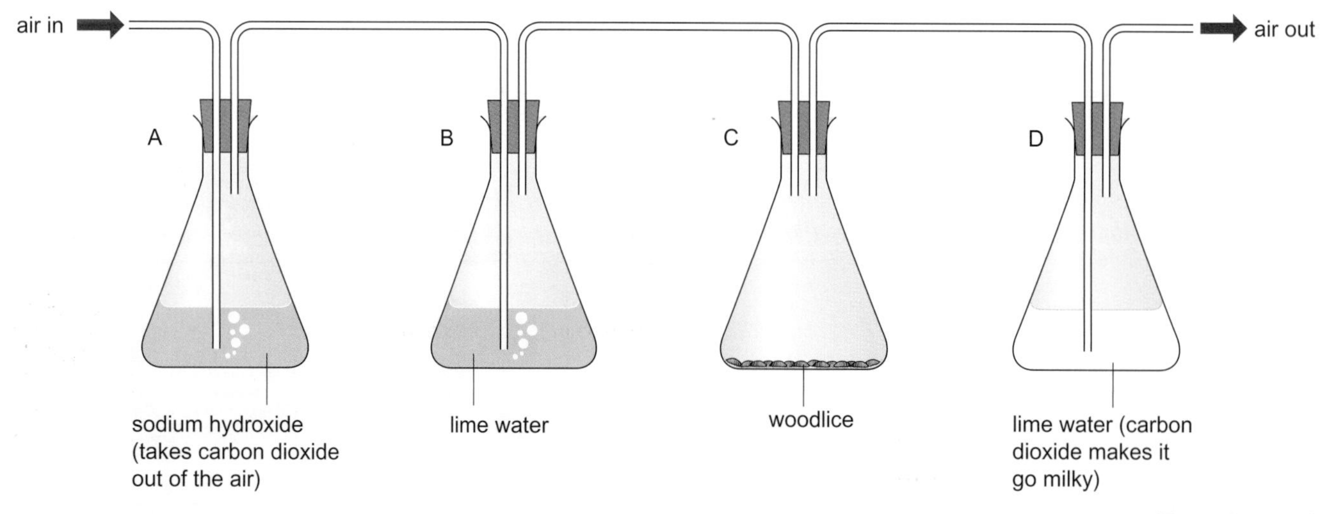

Experiment 2

Question 1 2 3

In experiment 2, the lime water in

- flask B shows that the air reaching the woodlice has no carbon dioxide in it
- flask D shows that the air has carbon dioxide in it.

We need flask B, to show that the carbon dioxide in flask D can have come only from the woodlice and anything living in or on them.

In experiment 1, we use

- one tube with woodlice in
- one tube without woodlice.

Then we can say that the woodlice cause any change.
We call the second tube the **control**.

Without the control, we can argue that something else might have caused the change. For example

- carbon dioxide could have leaked into the tube
- light, temperature or any other differences could have caused the change.

Question 4  5  6

Remember when you design experiments using living things

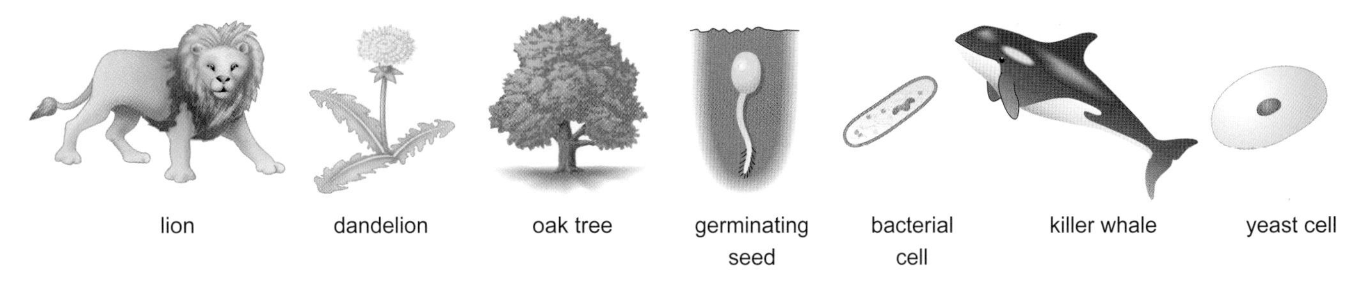

Remember that all living things respire.

- Use more than one living thing, because living things vary.
- Use a control, so that any change can be caused only by the variable that you are testing. It mustn't be caused by light, or temperature, or any other variable.
- Vary only one thing at a time.
- Treat living things with care and respect.

Review your work

Summary ➡

8B.1

1 Write down <u>one</u> food that provides energy for cells.

2 Write down <u>two</u> reasons why you need to make new cells.

3 Write down <u>one</u> time when your muscle cells need more glucose than normal.

4 People doing sports often use high-energy drinks. A slice of bread contains just as much energy as 200 cm^3 of the drink.
During exercise, the energy in the drink is more useful than the energy in the bread.
Suggest why.

5 What is respiration?

6 Write down:

  a <u>two</u> things that cells need for respiration;

  b <u>two</u> waste substances that are produced in respiration.

7 Draw a flow diagram to show how oxygen gets from the air to the cells in the body.

8 Why do all the tissues in the body need blood vessels near them?

9 Some parts of the body have a better blood supply than others. Explain why muscles need to have plenty of blood vessels.

10 **a** Write down <u>two</u> materials that diffuse from your blood to your cells.

  b Write down <u>two</u> materials that diffuse from your cells into your blood.

8B.2

1 Draw a time line to show how ideas about blood circulation changed.

2 Harvey did not find capillaries. Suggest why.

3 What technology was needed before capillaries could be discovered?

4 Draw a simple flow chart to show what Harvey and Malpighi found.

5 In your group, discuss what scientists have learnt from Harvey about studying the human body.

continued

6 Use books or the Internet to find out more about the work on blood circulation of one of:
- the Chinese
- the ancient Greeks (including scientists like Galen and Erasistratus)
- Islamic scientists (including scientists like Ibn-al-Nafis, also known as Al-Quarashi).

Agree in your group how to present the information.

7 Substances pass in and out of capillaries easily. Suggest why.

8 There are blood capillaries close to all your cells. Why is this?

9 What is your heart mainly made from?

10 Look at the double circulation diagram.
Write down, in order, the parts that the blood goes through.
Start and finish at the right-hand side of the heart.

11 How many times does blood pass through your heart each time it does a full circuit of your body?

12 The wall of the left side of the heart is thicker than the wall of the right side.
Suggest why.

8B.3

1 Write down the word equation for respiration.

2 Suggest the effect on the amount of energy released when cells don't get enough glucose and oxygen.

3 Write down <u>three</u> symptoms of altitude sickness.

4 The climber's cells are not getting enough oxygen.
Explain why this makes him feel tired.

5 If the amount of oxygen in the air drops slightly, pilots notice that they can't make decisions or concentrate.
Suggest why.

6 Jack found it hard to breathe. Explain why.

7 There are many jobs in which people should wear dust masks.
Suggest <u>two</u> of these jobs.

8 People with damaged lung tissue easily get out of breath.
Explain why.

9 Find out about the cause and symptoms of <u>one</u> other lung condition.
Make brief notes.
Remember to write down where you found your information.

8B.4

1 Look at the diagram.
Draw a flow chart to show the route air takes to get into your lungs.

2 What happens to oxygen and carbon dioxide during gas exchange?

3 Rapid gas exchange in the lungs is important. Suggest why.

4 What gives lungs their large surface area?

5 Why does a large surface area help gas exchange happen quickly?

6 Why do thin walls help gas exchange happen quickly?

7 Why can substances pass in and out of capillaries easily?

8 Having a lot of capillaries around the alveoli helps gas exchange.
Explain how.

9 Smoking affects the amount of oxygen that reaches your cells.
Write down two reasons.

10 Sports coaches ask the people that they train not to smoke.
Suggest why not smoking helps

a their sporting performance

b their health.

8B.HSW

1 What does the term collaborate mean?

2 List four risk factors for lung diseases.

3 Discuss these questions in your group.
In the fight against COPD in the UK,

a which agencies need to collaborate?

b what kinds of specialist need to be involved?

4 Out of 100 people who die next year, how many are likely to die from COPD?

5 What percentage of people in the better-off countries die from COPD?

6 What is the main risk factor for lung disease?

7 It will take many years to reduce the number of deaths from lung disease.
Suggest why.

continued

8 In the UK, 7% of sickness-related days off work are due to COPD. This is not just bad for the people who are ill.

In your group, discuss some of the effects of people's absence from work

- **a** on their families
- **b** on other people in their workplace
- **c** on the country.

8B.5

1 What is the liquid on the mirror?

2 Where does it come from?

3 Why does water vapour show up in exhaled air on a cold day?

4 What are the two waste products of aerobic respiration?

5 **a** Which contains more oxygen – inhaled or exhaled air?

b Explain why this is.

6 Which gas is there more of in exhaled air than inhaled air?

7 Where is this extra gas made?

8 The amount of water vapour in inhaled air varies. Why is this?

9 The runner in the photograph is making more carbon dioxide than he does when he is asleep. Suggest why.

8B.6

1 **a** Write down two substances that you can use to detect carbon dioxide.

b Describe the change to each substance when carbon dioxide is present.

2 In experiment 2, the woodlice breathe in air that has no carbon dioxide in it. How do you know this?

3 The change in the lime water shows that there is carbon dioxide in the air going through flask D. Where did it come from?

4 Describe the results of experiment 1.

5 What can you conclude from experiment 1?

6 Katie did an experiment like this with maggots. But she didn't use a second tube. The indicator changed colour. Her teacher told her that she couldn't conclude that maggots produced carbon dioxide. Why is this?

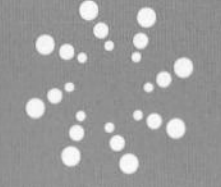

Scientific enquiry

8C.1 Micro-organisms and how to grow them

You should already know | Outcomes | Keywords

Types of micro-organism

Some living things are so small that we can only see them through a microscope. We call them **micro-organisms** or microbes.

They include **viruses** and **bacteria**, and some **fungi**.

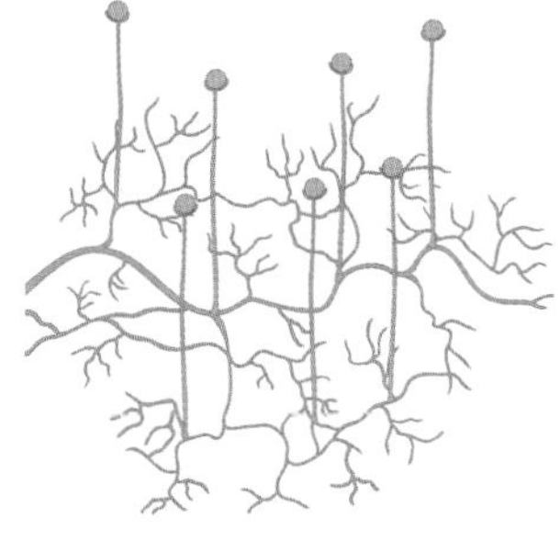

Moulds are thread-like fungi. Some cause decay. We use others for making antibiotics.

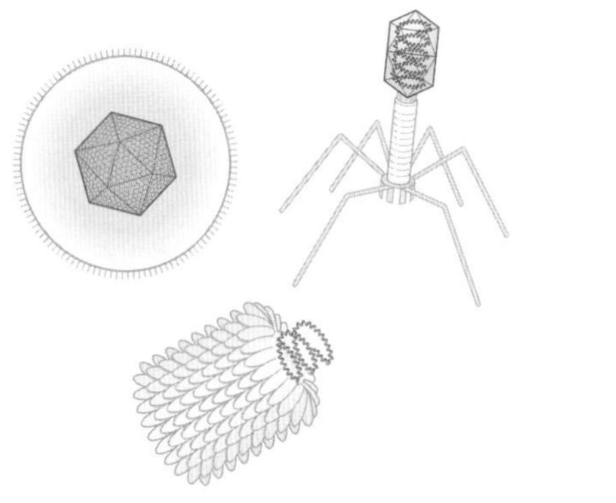

Examples of viruses

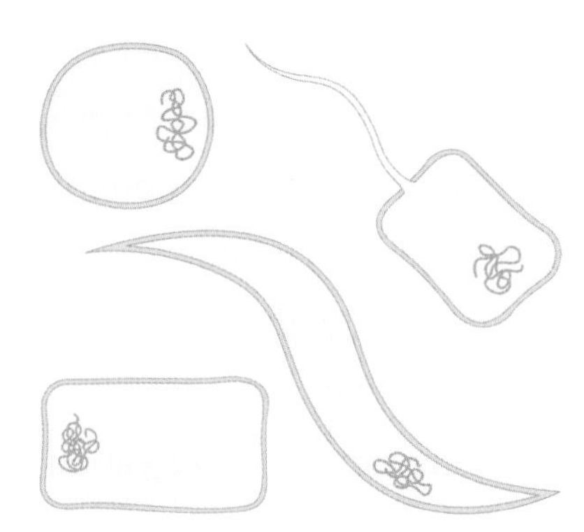

Examples of bacteria

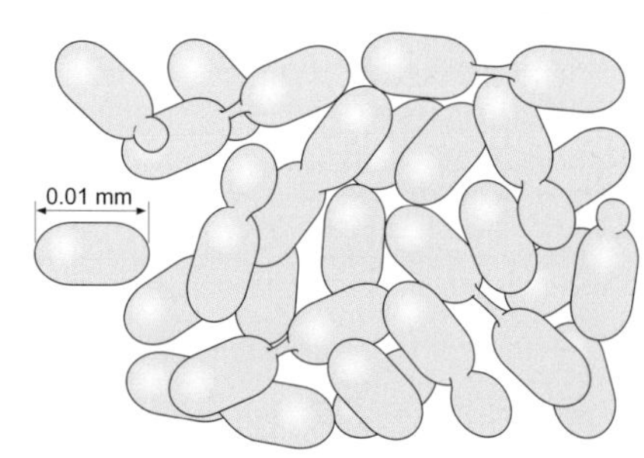

Yeast is a fungus made of single cells. It reproduces by budding off new cells. We use it for making bread and wine.

Question 1

Although micro-organisms are very small, they have a huge effect on our lives. Some are very useful. Others cause **disease**.

Question 2

Here are two micro-organism fact files.

FACT FILE: Viruses

Average size	0.0001 mm
Structure	A strand of genetic material wrapped in a protein coat.
Found	Viruses can reproduce only inside living cells.
Uses	To kill pest animals.
Diseases	Common cold, influenza (flu), measles, AIDS, yellow fever, rabies. Viruses cause disease in animals, plants and even other micro-organisms.

FACT FILE: Bacteria

Average size	0.001 mm
Structure	Bacteria are single-celled with a strong cell wall. Their genetic material is not in a nucleus.
Found	Most bacteria live in water, soil and decaying matter.
Uses	To make yoghurt, cheese and vinegar.
Diseases	Typhoid, cholera, food poisoning.

Question 3  4  5  6

How to grow micro-organisms

Yeast is the most commonly grown micro-organism in the world.
We use it to make

- beer
- wine
- bread.

Yeast is a living thing, so it breaks down food in respiration to get energy.

If it uses oxygen, we call it aerobic respiration.

But yeast can also respire without oxygen. When it does this, it makes alcohol.

In both cases, it produces carbon dioxide.

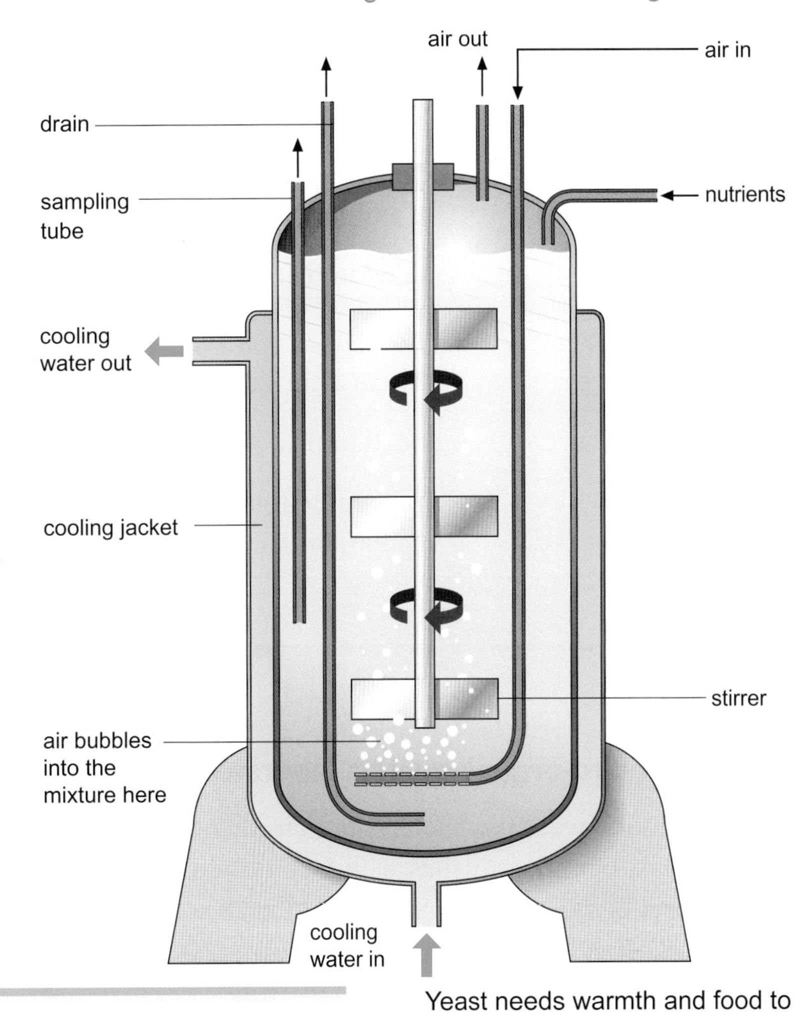

Yeast needs warmth and food to grow. This is how yeast is grown on a large scale. Each vat is big enough to fit a family car inside!

Question 7 **8**

Using yeast to make bread

To make bread dough, you use yeast, flour, water and a little sugar and salt. Laura did an experiment to see if sugar helps dough rise. She measured the amount of dough in each measuring cylinder after 30 minutes.

30 minutes

bread dough with sugar

Amount of sugar in the dough in g	Volume of the dough after 30 minutes in cm^3
0	40
5	52
10	66
15	74
20	80

Question 9  **10** **11**

Yeast is just one type of fungus that we use.
There are lots of others.

Product made by fungi	Use
penicillin	an antibiotic to treat some diseases
citric acid	added to make soft drinks taste tangy
cortisone	to treat arthritis
pectinase	added to fruit juice to make it clear
mycoprotein	a substitute for meat that is suitable for vegetarians (for example, Quorn®)

Question 12

Growing micro-organisms in a laboratory

In a laboratory, we grow bacteria in Petri dishes – small plastic or glass dishes with lids. Food for the bacteria is mixed with a jelly called agar. Each type of micro-organism needs its own particular mix of minerals and food to grow.

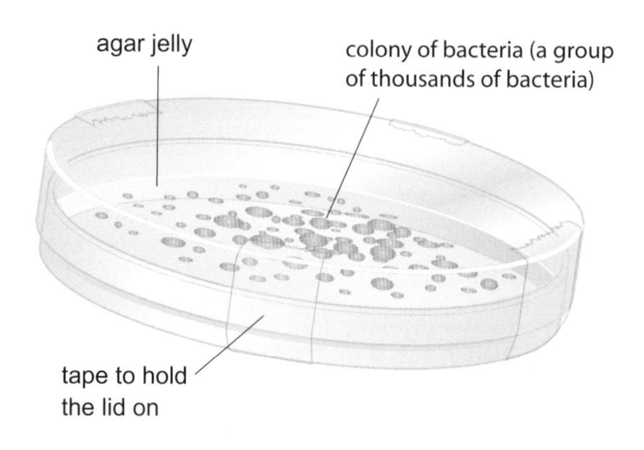

Lids are always taped onto Petri dishes.

Question 13

Everything we use to grow micro-organisms must be very clean. We have to sterilise Petri dishes and agar by heating them to over 100 °C. We must keep all benches and equipment clean.

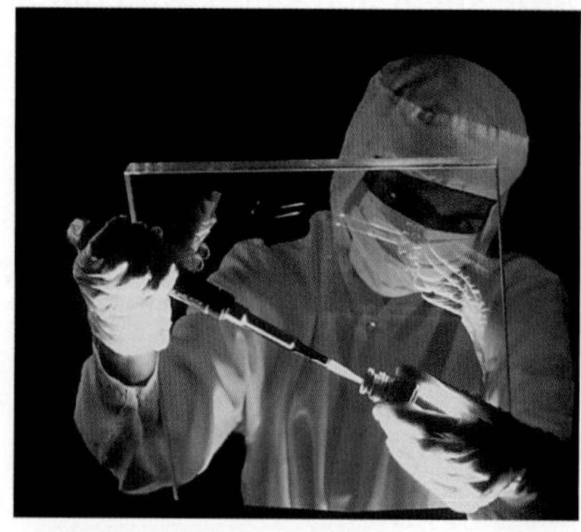

A scientist who works with micro-organisms is called a microbiologist.

Question 14 **15**

You should already know

Outcomes

Keywords

The spread of infection

Illnesses caused by micro-organisms are called **infections**, or infectious diseases.

Diseases can spread in many ways.

- **By animal** – e.g. malaria and yellow fever by mosquitoes, and the rabies virus by mammals.

- **By droplet** – e.g. tuberculosis (TB) bacteria and chickenpox virus.

- **In food** – e.g. *Salmonella* bacteria from flies, dirty hands or dirty knives and dishes.

- **In contaminated water** – e.g. typhoid bacteria.

- **By direct human-to-human contact**.
  - Athlete's foot is a fungal disease spread by touch.
  - HIV/AIDS is spread when people have sex and don't use a condom.
  - A pregnant woman can pass on a disease to her unborn child.
  - A mother can pass on a disease to a baby through breast milk.

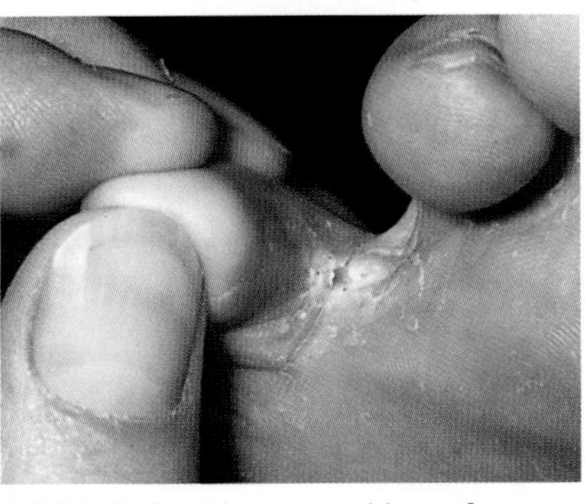

Athlete's foot is caused by a fungus.

Question 1    

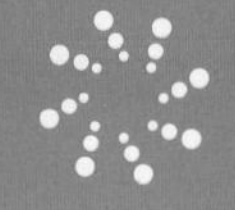

8C.3 The evidence that stopped an epidemic (HSW)

You should already know | Outcomes | Keywords

How to stop an epidemic

An **epidemic** is an outbreak of a disease that affects a lot of people.

In 1854, there was a terrible outbreak of a disease called cholera in Soho (an area of London). Cholera was a common disease in polluted city areas, so people thought that you caught cholera from bad air.

Look at the table to see how quickly the disease spread.

Date in 1854	Number of new cases of cholera	Number of deaths from cholera
31st August	56	3
1st September	143	70
2nd September	116	127

Question 1 | 2

Many more people would have died had it not been for a doctor called John Snow. He thought that infected water – not bad air – caused cholera.

He heard about the deaths and went to investigate.

At that time, most people got their water from public street pumps.

John Snow suspected that people in Soho caught the **infection** from water from their pump in Broad Street.

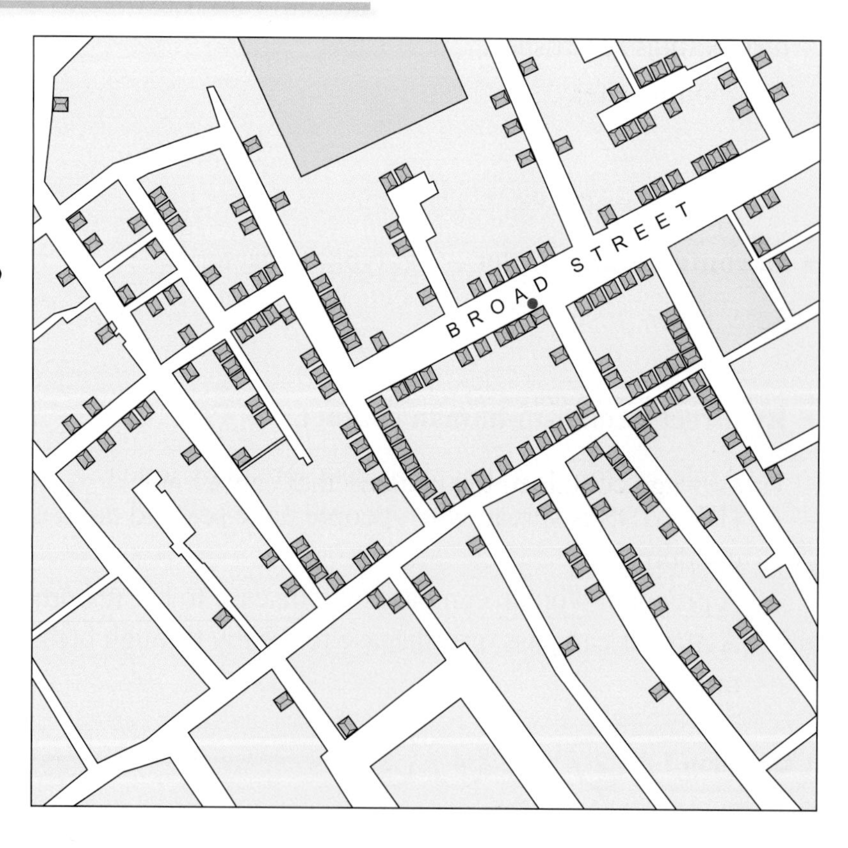

Key
- pump
- cholera death
- workhouse
- brewery

John Snow's evidence

- Most people who died lived near the Broad Street pump.
- Only 5 out of 500 people died at the workhouse round the corner from the pump. The workhouse had its own well.
- No one in the local brewery died. All the workers drank beer instead of water.

- Two ladies who lived 5 miles away died of cholera. They had a bottle of water brought to them from the Broad Street pump because they liked the taste.

By 7th September, three-quarters of the people living in Soho had left. Many of those who remained were ill and 28 more died that day.

John Snow showed the Parish Board his evidence. They agreed to remove the handle of the pump the next day. People could no longer drink the water. The number of new cases of cholera began to fall.

What happened?

A few months later, John Snow found the cause of the epidemic.

- Water for the pump came from an underground well.
- Number 40 Broad Street had an underground cesspit for sewage.
- During August, a baby at number 40 was ill with cholera. After his mother washed his nappies, she tipped the water into the cesspit.
- The cesspit wall was cracked and the sewage leaked out into the Broad Street well.
- The sewage contained the bacteria that cause cholera.

Question 4

Check your progress

8C.4 Protecting ourselves against disease (HSW)

You should already know | Outcomes | Keywords

Your body's defences

Your body can defend itself against micro-organisms.

Look at the diagram.

Each time people cough or sneeze, they spray little droplets into the air.

- Someone with tuberculosis (TB) coughs out TB **bacteria**. TB is a serious **disease** that destroys lung tissue.
- Droplets sneezed out by people with colds contain thousands of **viruses**.

If you are nearby, you can breathe them in. Sometimes they get past the defences in your air passages and into your lungs.

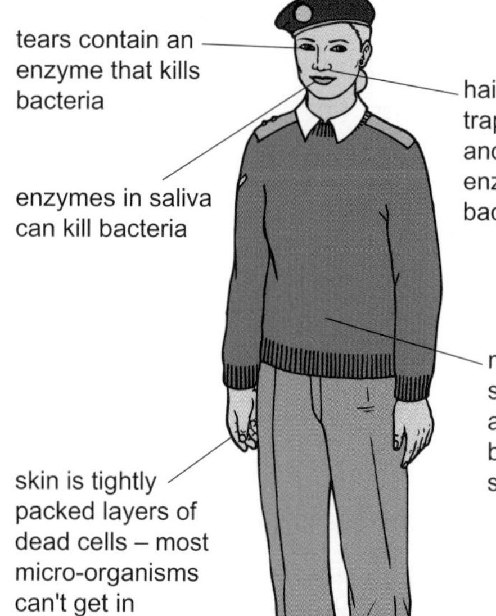

How your body defends itself against micro-organisms. Unfortunately, micro-organisms get past these defences.

Question 1 2 3 4

Antibiotics help to fight some diseases

Antibiotics are substances made by living things. They kill bacteria but not viruses.

In the past, some people used moulds on wounds to heal them. Scientists saw that some moulds stop bacteria growing.

Question 5

In 1928, Alexander Fleming noticed that there were no bacteria where a mould grew on one of his dishes of bacteria. Moulds are **fungi**. The mould had made a substance that stopped the bacteria growing. He called the substance **penicillin** after the mould *Penicillium*. Doctors started to use it in 1942. Since then, penicillin has saved millions of lives.

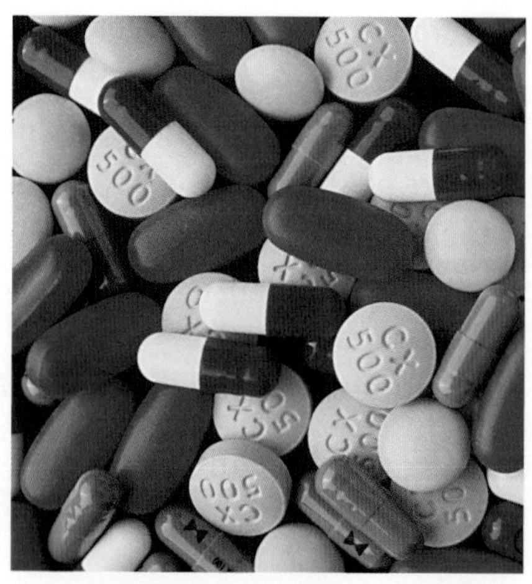

Penicillin kills only some kinds of bacteria. We now have lots of different antibiotics to treat different infections.

Question 6 7 8 9 10 11

How your body destroys micro-organisms

Your blood contains two different types of **white blood cells**. One type engulfs (traps) micro-organisms and destroys them. The other type makes substances called **antibodies**.

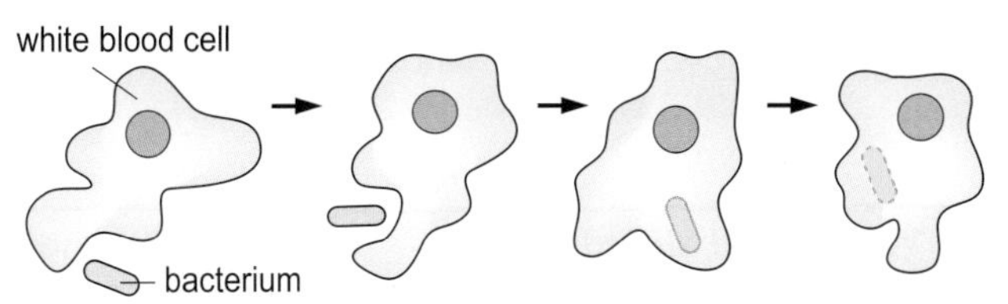

Some white blood cells take in micro-organisms and destroy them.

- Each antibody acts against one type of micro-organism only. So you need different antibodies against different micro-organisms.
- It takes time for your body to make new kinds of antibodies. So, you feel ill until you have made enough of the right antibodies.
- When you have a disease, your white blood cells make antibodies against it.
- In a second attack, antibodies can destroy the micro-organisms before they make you ill. So you are immune to the disease. You have **immunity**.

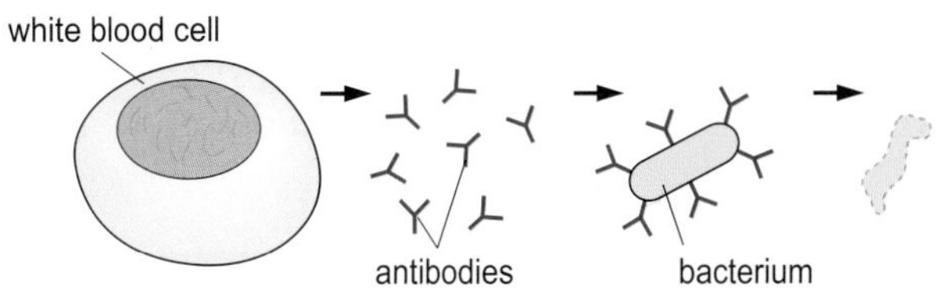

Some white blood cells make antibodies that stop micro-organisms working properly.

Question 12 

Jabs can also make you immune to a particular disease

A 'jab' is an injection of a **vaccine**. This makes you immune to the disease. So we call having a jab **immunisation**.

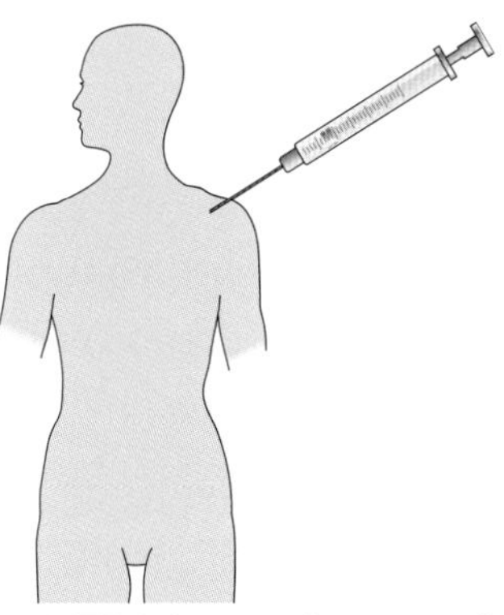

A TB jab contains a weak version of TB bacteria. After the jab, some white blood cells make the right antibodies to kill TB bacteria.

Question 15

Before scientists developed the TB jab, many people died of TB.

Those who had the disease and survived were then immune.

The TB vaccine contains a weakened form of TB bacteria. So it has the same effect on your immune system as being infected with TB but it doesn't make you ill. At worst, your arm may feel sore.

If TB bacteria get into your body in future, you quickly make antibodies to destroy them. So you will not be ill.

Colds and flu are different. The viruses that cause them change all the time. Your white blood cells don't recognise the new forms. So you can get these infections again.

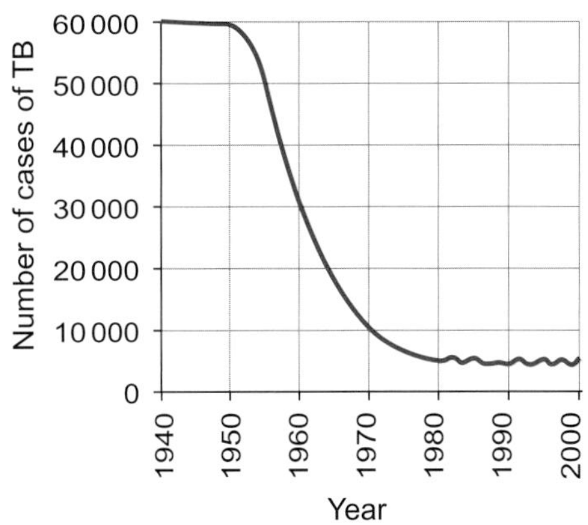

Total annual number of cases of TB in England and Wales from 1940 to 2000. (Data from the Public Health Laboratory Service.)

Question 16 

20

Review your work

Summary ➡

8C.HSW Maggots and other medical marvels

You should already know

Outcomes

Keywords

In the past, doctors treated sick people with things such as mouldy oranges, leeches and maggots. All of these probably worked some of the time. Then doctors found different treatments.

Question 1

Leeches – past and present

Leeches suck blood. In ancient Egypt, Greece and Rome, physicians thought that excess blood caused some illnesses. They used leeches to suck this excess blood from their patients. This is called blood-letting.

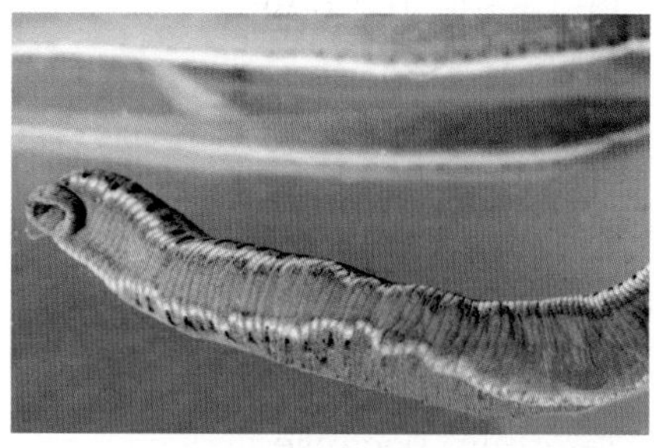

Doctors kept leeches in jars. Sometimes, their patients recovered after blood-letting. Sometimes, they died.

Treatment with leeches continued for centuries. Doctors were even known as 'leeches'. William Harvey, who discovered blood circulation, used them.

In the 1860s, London hospitals used about 7 million leeches a year.

By the 1930s, most hospitals had stopped using them.

Question 2 3

Now, doctors know why leeches helped some patients. So they are using leeches again. Leech saliva contains an anaesthetic – so their bite doesn't hurt.

Leech saliva also contains substances that

- stop blood clotting
- dissolve blood clots.

Doctors are making use of these properties. Look at the Fact file.

Fact file

Now doctors use leeches to

- thin the blood to prevent clotting after heart surgery
- dissolve blood clots – in major blood vessels, clots can kill
- unblock tiny blood vessels following microsurgery such as sewing on chopped-off fingers
- reduce bruising after surgery.

Question 4 5 6

In the past, doctors used leeches to treat infections. But there is nothing in the Fact file to suggest that this treatment could have worked.

Maggots are back in fashion

Surgeons in Napoleon's armies noticed blowflies sometimes laid eggs in wounds. The eggs hatched into maggots that fed on the damaged tissues. The wounds healed more quickly and the soldiers were less likely to die from infection.

During the American Civil War, surgeons actually put maggots into wounds. They didn't know how they worked.

Later, doctors used antibiotics to treat infections instead.

adult fly

eggs laid in wound

maggots (fly larvae) feed on decaying flesh

pupa

Question 7 **8**

Researchers decided to look again at the evidence about maggots. They found out how they worked.

So some doctors are using maggots again. 200 maggots can clean a wound or an ulcer in about 5 days. Normally, ulcers can take months to heal.

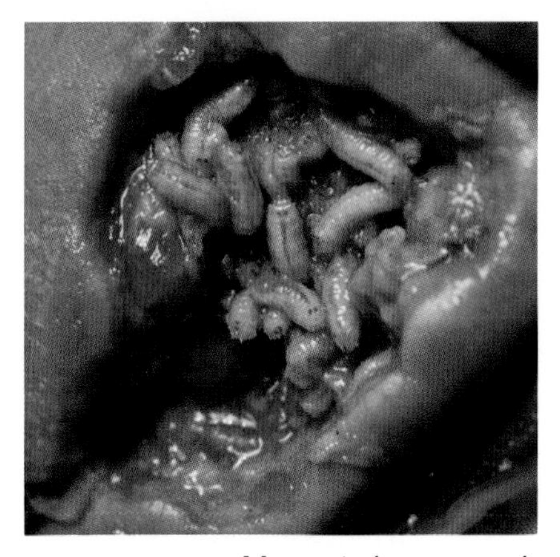

Maggots in a wound.

Doctors are also hopeful about maggots as a treatment for wounds infected with MRSA. This bacterium is sometimes called a 'superbug' because

- it is resistant to many antibiotics
- it kills about one in five infected people.

Question 9

Rediscovering phages

When scientists discovered **bacteriophages** (also called **phages**), they thought that they were future cures for bacterial infections.

Russian doctors found a phage that killed the bacteria that cause gangrene. During the Second World War, this saved the lives of soldiers with gangrene. Russian doctors still use phages to treat sore throats.

In the West, scientists thought that antibiotics were the future. But many bacteria can survive antibiotics. We say they are **antibiotic resistant**. So now they are looking again at uses of phages, including

- killing MRSA inside the nose so that it doesn't get into wounds
- spraying meat to kill the bacteria that cause food poisoning.

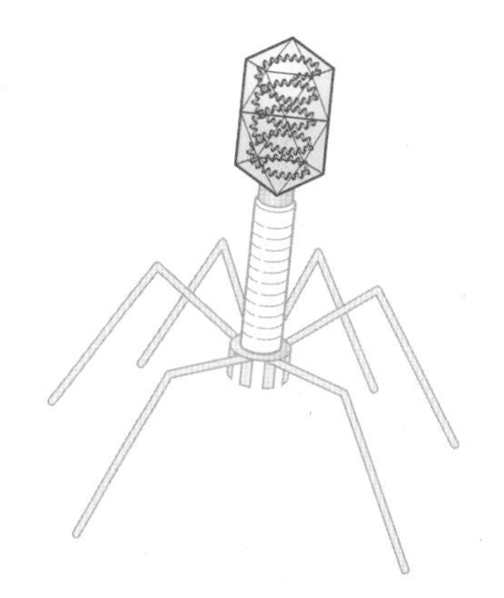

Bacteriophages are **viruses** that kill bacteria. But each kind of phage kills only one kind of bacterium.

Question 10 **11**

8C.1

1 Write down <u>three</u> different types of micro-organism.

2 Write down <u>two</u> things we make using micro-organisms.

3 How many times bigger is the average bacterium than the average virus?

4 Which type of micro-organism is the smallest?

5 Which <u>two</u> types of micro-organism make food rot?

6 Draw a table with three columns. Write in the three headings Bacteria, Fungi and Viruses, like this.

Bacteria	Fungi	Viruses

Use information from the Fact files to complete the table.

7 Write down <u>three</u> things that both yeast and humans do.

8 Look at the diagram of the vat.
How can you tell from the design that the yeast cells in it respire aerobically (using oxygen)?

9 What is the effect of the amount of sugar on the height of the dough?

10 What is the gas that makes dough rise?

11 When making bread, you leave the dough to rise for several hours.
Then you bake it for 30 minutes at 200 °C.
What do you think happens to yeast during baking? (Remember, yeast is a living thing.)

12 Look at the table.
Write down <u>two</u> medicines made by fungi.

13 Bacteria are so small that you can only see them with a microscope.
Why can you see bacteria growing on agar?

14 Look at the photograph on the right.
Write down <u>three</u> things the worker is wearing so that unwanted micro-organisms do not contaminate his work.

15 Sometimes unwanted micro-organisms grow in dishes.
Suggest <u>one</u> way this can happen.

8C.2

1. A child at a birthday party can give the other children chickenpox.
   Suggest how.
2. You should wash your hands before you prepare food.
   Explain why.
3. Flies can spread food poisoning.
   Suggest two ways of preventing this.
4. Write down one way of making water safe to drink.
5. Write down two ways that a mother can pass on a disease to her baby.

8C.3

1. How many people died from cholera in the first three days of the epidemic?
2. Suggest who could have collected the data that alerted John Snow.
3. Imagine you are the person who took the handle off the pump.
   A crowd of local people complain because they now have to walk 10 minutes to get drinking water.
   In your group, discuss what to say to them.
4. John Snow could not see the bacteria that cause cholera.
   Suggest why he was so sure that there were bacteria in the well at Broad Street.

8C.4

1. Micro-organisms don't normally get into your body through your skin.
   Suggest why.
2. In which part of your body does acid kill micro-organisms?
3. Animals often lick their wounds.
   How does this help wounds to heal?
4. You are less likely to catch TB if you breathe through your nose rather than through your mouth.
   Suggest why.
5. Why should you cover your nose and mouth when you sneeze?
6. Would you put mouldy food on a cut to help it to heal?
   Discuss this in your group.

continued

7 What stops bacteria from growing close to *Penicillium*?

8 Normally, new drugs are tested and trialled for years before doctors can use them. Doctors used penicillin as soon as it could be made in large quantities.
Suggest why.
(Hint: the Second World War lasted from 1939 until 1945.)

9 Before penicillin was discovered, one in three people who caught pneumonia died. When penicillin was used, only one in twenty people died.
If 300 people caught pneumonia, how many would probably have died:

a before penicillin was discovered?

b after penicillin was discovered?

10 The discovery and development of penicillin is an interesting story.

a Find out more about the discovery of penicillin and about the work of Howard Florey and Ernst Chain.

b In your group, discuss what the story of penicillin tells us about the way scientific knowledge can develop.

11 Mrs Sharples has flu (caused by a virus). She wants antibiotics to make her better.
Explain why you will not give her any antibiotics.

12 You don't usually catch the same disease twice.
Explain why.

13 When you have had a cold, you become immune to that particular virus. A few months later, you can catch another cold.
Explain why.

14 Breast milk contains the mother's antibodies.
How does this stop a breast-fed baby from catching some diseases?

15 Find out what vaccinations you have had and when you had them.

16 Your white blood cells make antibodies in response to the TB vaccine.
Explain why.

17 Copy and complete this flow chart to show how TB immunisation stops you catching TB.

18 Look at the graph.
In what year do you think TB jabs were first given to nearly all children in England and Wales?

continued

19 In the year 2000, there were still over 6000 cases of TB in England and Wales. Not everyone is immunised against TB.
Write down two possible reasons why someone might not have had the TB jab.

20 Vaccination doesn't just protect you. If most people are vaccinated, the whole population is protected against an outbreak.
Suggest why.

8C.HSW

1 In your group, take a vote on whether you are willing to be treated using leeches and maggots.

2 Too much blood-letting makes people more ill.
In your group, discuss a possible explanation.

3 Suggest why doctors stopped using leeches.

4 A surgeon sewed a patient's thumb back on. She asked the patient if she could put a leech on the end of the thumb. She said that research showed that this makes surgery more likely to work.

a What sort of evidence might the researchers have collected?

b Suggest how this use of leeches helps.

5 Dan had a blood clot in a vein in his leg. The clot might move to his heart or lungs and block important blood vessels. The consultant gave him an injection of a substance from leech saliva.
Suggest how leech saliva helped Dan.

6 In which war did doctors use antibiotics for the first time?
(Hint: look at Topic 8C.4.)

7 Some people don't want their wounds to be treated using maggots.
Suggest why.

8 Working as a group, suggest some advantages and disadvantages of using maggots on infected wounds.

9 In your group, take another vote on whether you are willing to be treated using leeches and maggots. If anyone has changed their mind, find out why.

10 Scientists in the West reduced the amount of research on the use of phages to treat bacterial infections.
Suggest why.

11 Scientists in the West are now looking again at using phages against bacteria.
Suggest why.

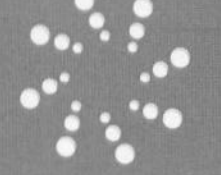

8D.1 Classifying animals and plants

Scientific enquiry

You should already know

Outcomes

Keywords

Living things are adapted to the habitats in which they live.

When you study a **habitat**, you need to be able to identify the plants and animals that you find. It helps if you know which group they belong to.

When we put plants and animals into groups we say that we **classify** them.

Classifying animals

You should already know that we classify animals into

- animals with backbones, called **vertebrates**
- animals without backbones, called **invertebrates**.

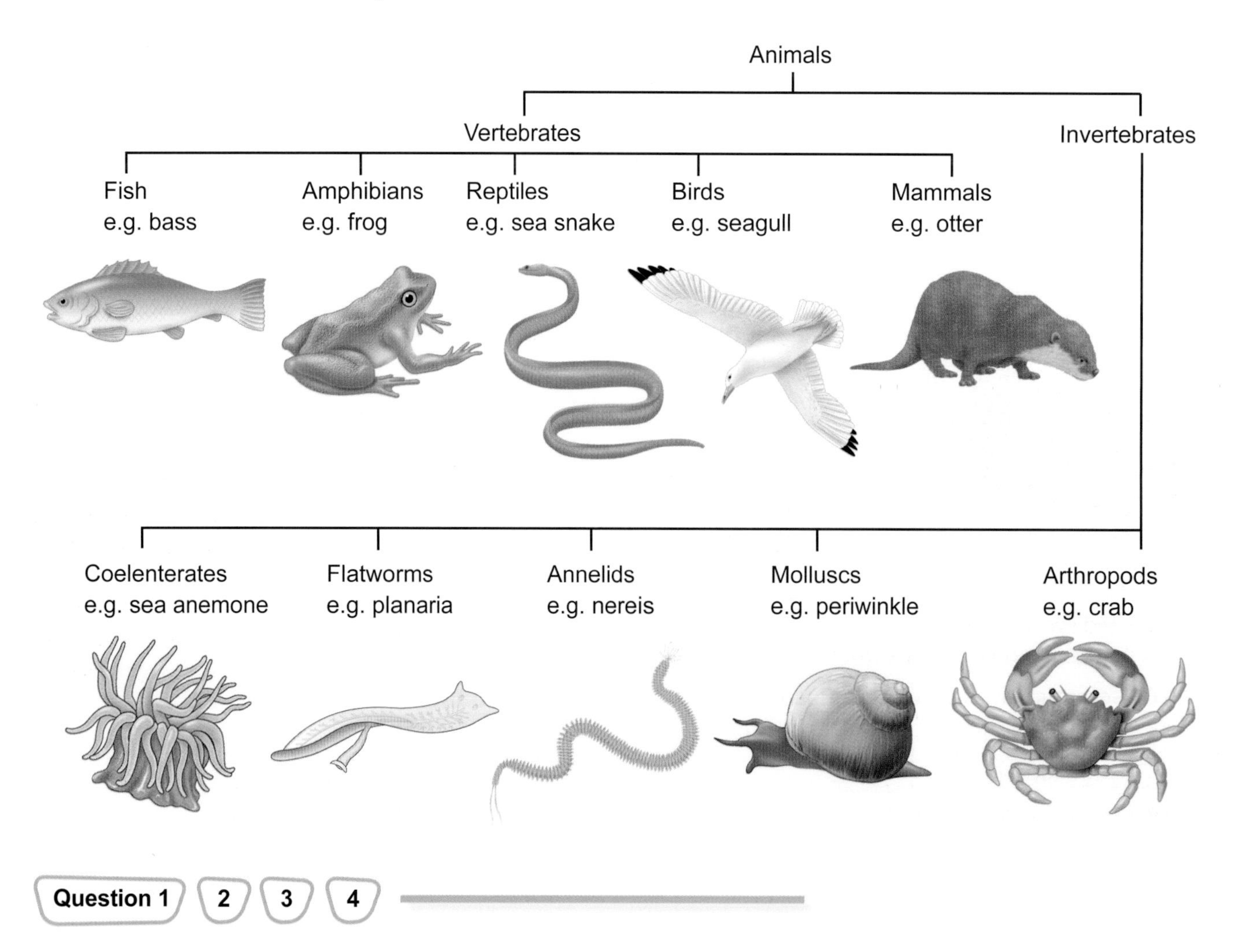

Question 1 2 3 4

Now let's classify green plants

There are so many plants that we divide them into groups to help us to identify and to study them.

There is more than one way of doing this.

Some plants have a special transport system for food and water called a vascular system. So we call these plants **vascular plants**.

We call plants without a special transport system **non-vascular plants**.

We divide both these groups of plants into smaller groups.

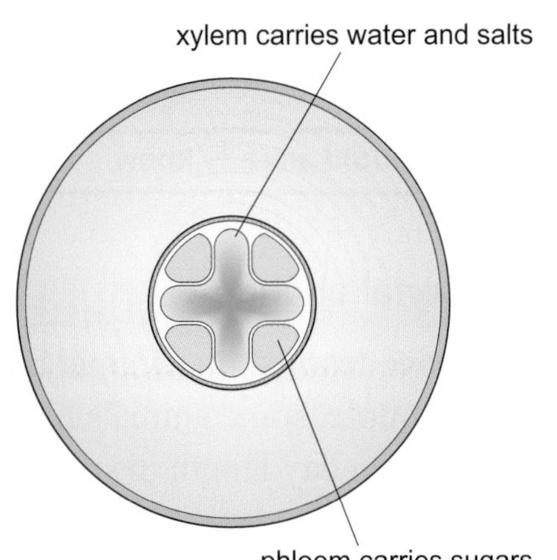

A slice through the root of a vascular plant.

Question 5 | 6

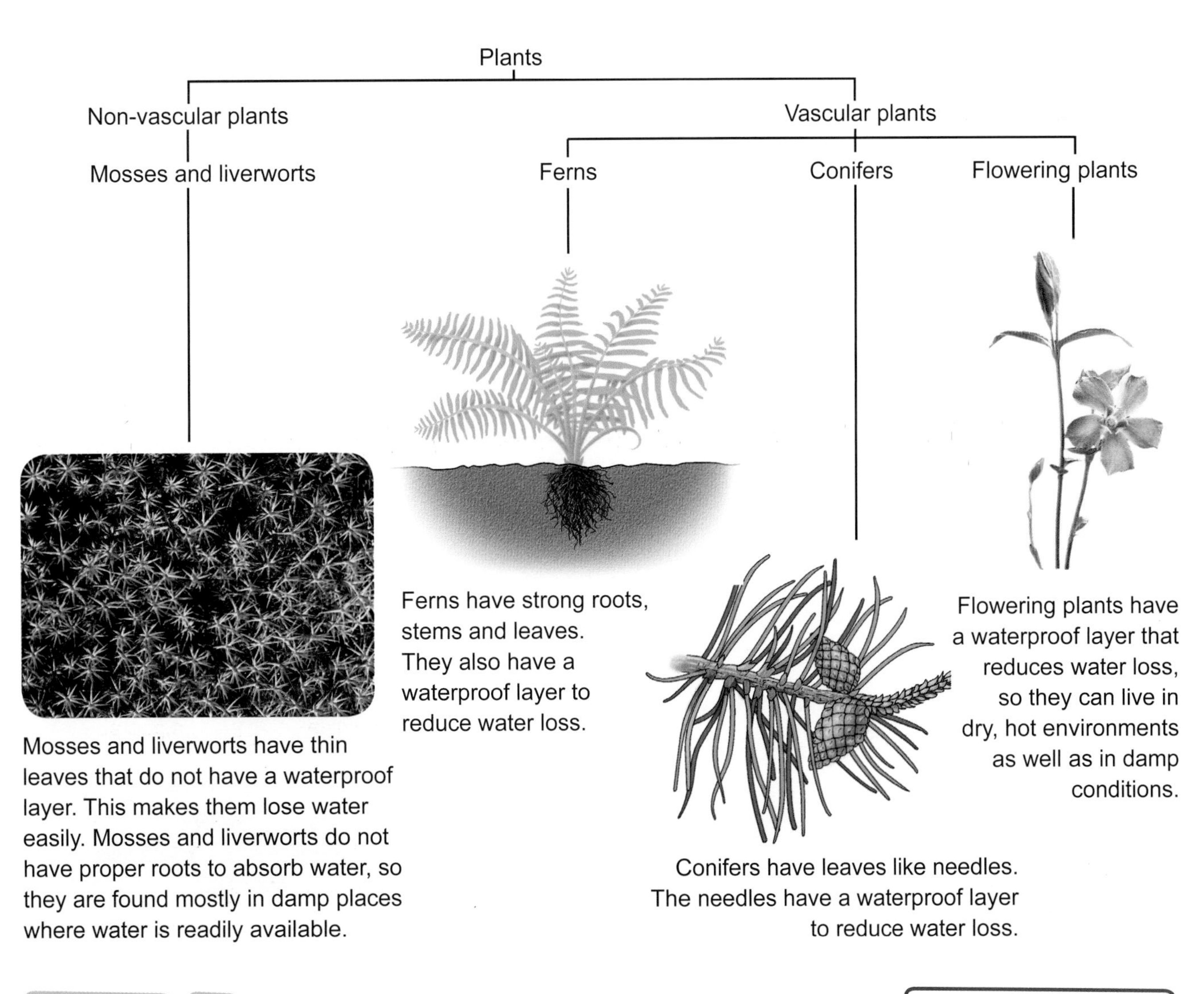

Mosses and liverworts have thin leaves that do not have a waterproof layer. This makes them lose water easily. Mosses and liverworts do not have proper roots to absorb water, so they are found mostly in damp places where water is readily available.

Ferns have strong roots, stems and leaves. They also have a waterproof layer to reduce water loss.

Conifers have leaves like needles. The needles have a waterproof layer to reduce water loss.

Flowering plants have a waterproof layer that reduces water loss, so they can live in dry, hot environments as well as in damp conditions.

Question 7 | 8

Check your progress

8D.2 Behaviour for survival

You should already know

Outcomes

Keywords

Seasonal change

You have learned that animal behaviour changes with the seasons. For example, some animals hibernate and some migrate. Changes in day length in spring and autumn cause these changes in behaviour.

Breeding seasons

Most animals breed only when there is plenty of food. So many animals in the UK breed in the spring. When the day length is right for them to breed, animals make more **sex hormones**.

Remember

The patterns of behaviour that newborn animals show are **innate**, or **instinctive**.

For example, newborn babies will grip your finger and move their heads in search of a nipple. Later, they learn new behaviours.

Sex hormones stimulate **breeding behaviour** such as courting and nest building. These are innate behaviours. But animals learn by trial and error to do these things better. This is **learned behaviour**. Animals that have learned to do better are the most likely to attract a mate.

Day length: external factor

Hormones: internal factor

Question 1 | 2 | 3 | 4

Tides affect when some marine animals breed.

Some marine animals release eggs and sperm into the sea. To make sure that the eggs are fertilised, they all need to do it at the same time.

Horseshoe crabs gather on the shore to lay their eggs. There are so many eggs and young that predators can't eat them all.

Horseshoe crabs and turtles lay eggs in pits on sandy shores. Female turtles lay theirs above the high-tide level. They leave the water at night during the highest tides. Tides are highest at the new moon and the full moon.

Male horseshoe crabs hold onto females and fertilise the eggs as they lay them.

Question 5 | 

Breeding behaviour

Species that breed in different places have different **behaviour patterns**. These include patterns of mating and parental care.

Gulls swallow fish and then bring them back to feed their young. Later, their young learn to fly and to find food.

Look at the table.

Black headed gull

Kittiwake

The breeding behaviour of these two seagulls is very different.

	Black-headed gull	Kittiwake
Nest	in colonies on the ground – a simple hollow	on small cliff ledges – use mud and seaweed to make a nest like a cup
Courting and mating	male displays to female and gives her food	male displays to female and gives her food
Alarm calls	often	rare
Mobbing predators	yes	no
Camouflaged young	yes	no
Feeding chicks on regurgitated food	parent has red bill young feed from bill	parent red inside mouth young feed from throat
When threatened by an adult	young run away	young stay still and hide their beaks

Question 7  8  9  10

Newly hatched young of geese and ducks are better developed.

- They instinctively recognise and follow their mother after they hatch. We say that they become imprinted on her. But they can become **imprinted** on the first moving object they see. Look at the picture.
- They leave the nest after hatching but their mothers protect them.
- They can feed themselves.

Ducklings imprinted on Konrad Lorenz (1903–1989), who researched this behaviour.

Question 11  12  13

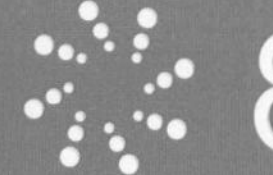

8D.3 Interactions in a habitat (HSW)

You should already know

Outcomes

Keywords

Collecting data to answer questions about a habitat

Molly and her class did some fieldwork on a rocky shore.

There are lots of plants and animals there. Animals are harder to see in many other habitats.

The plants and animals found on the shore depend on each other. They are called a **community**.

Question 1

Asking questions before the visit

In the lesson before the visit, Molly's teacher asked the class to think about what they'd like to find out about the rocky shore.

What plants and animals live on the rocky shore?

Will we find the same plants and animals on the same part of the shore?

How can we find the numbers of the different plants and animals?

Will different areas of the shore have different environmental conditions?

Question 2

It is often not possible to count all the individuals in a **population**, so we take a **sample**. Then we estimate the total number.

Question 3

It is easier to study the shore when the tide is out, but the seaweed makes it slippery.

When the tide comes in, it brings with it tiny floating plants and animals called plankton. Many shore animals feed on plankton.

Collecting the information

Molly decided to see if different plants and animals lived on different parts of the shore.

Her teacher showed her how to use a **quadrat** to sample the shore life. She

- threw a plastic card on the ground
- placed the quadrat so that the card was in the centre
- wrote down the names of the different plants and animals in the quadrat.

	Quadrat number				
Name	1	2	3		
Periwinkle	III	II	卌		
Anemone	卌	III	I		
Barnacle	IIII	I			

Molly suggested taking these samples:

- ten near the top of the shore
- ten half-way down the shore
- ten near the sea.

Molly's teacher thought that this was a good idea. She asked David's group to do the same experiment so that they could compare the results when they got back to school.

Question 4

Looking at the results

When Molly looked at her results, she could see that different plants and animals lived on different parts of the shore. Her teacher asked her to call the areas she sampled the upper, middle and lower shore. She explained that the habitats are different, and different habitats support different living things. Look at the pictures.

The upper shore spends a lot of time out of the water. It is exposed to different weather conditions.

The middle shore spends less time under water than the lower shore. It is not exposed to the weather for as long as the upper shore.

The lower shore spends a lot of time under the water.

Question 5

From results to conclusions

Molly summarised her results in a table.

Area	Species found
Upper shore	Most of the rock is covered by black lichen. In rock crevices I found a few tiny periwinkles and barnacles.
Middle shore	I found lots of barnacles, limpets and mussels in this area of the shore. I also found some small pieces of seaweed in this area.
Lower shore	The seaweed was long and flexible. I could not remove it from the rock. I found crabs, starfish, limpets, fish and sea anemones here.

Question 6 

Explaining the differences

Back at school, the teacher asked the class to try to explain why there were different communities in different habitats. They thought that the environmental conditions were different in the different areas of the shore.

These are their ideas:

Upper shore

- This area can dry quickly. It spends a lot of the time not covered by the seawater. So there is less feeding time for the animals. It is often exposed to very hot or very cold weather conditions. Sometimes rain makes the water less salty.

Middle shore

- This area spends more time under water than the upper shore, but less time under water than the lower shore. The water brings with it a rich supply of food.

Lower shore

- This area spends most of the time under water. So there is less variation in temperature and not such a problem of drying.

Lichens are made of a green alga and a fungus. They often live on rocks and tree trunks. They can survive because they take a long time to dry out.

Question 8

Plants and animals on the shore have ways of making sure that the waves don't wash them away.

Many species of seaweed are long and flexible. They also have very strong holdfasts, which fix them to the rocks.

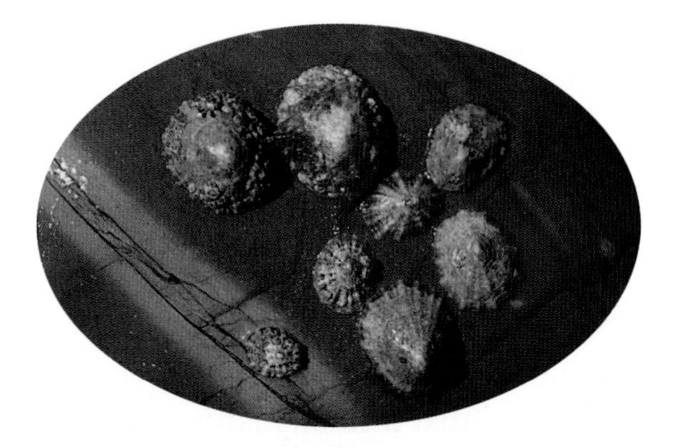

Limpet shells fit closely to the rock so that they don't dry out. They feed on tiny seaweeds on the rock when the tide is in.

Mussels live attached to rocks. They filter plankton from the water when the tide is in.

Barnacles are cemented to the rock. When the tide is in, they filter plankton from the water.

Question 9 10 11

Population size and environmental conditions

Molly's teacher agreed with their ideas. She said that the size of the population of an organism is affected by **environmental conditions**, such as the amount of light, water and nutrients.

The teacher also said that organisms have more chance of survival where there is less variation in temperature.

Molly looked at the results of her quadrats for the population of barnacles.

Area	Number of barnacles
Upper shore	5
Middle shore	62
Lower shore	0

Question 12 13 14 15

8D.4 How living things depend on each other

You should already know | Outcomes | Keywords

Food chains and webs

A **food chain** shows how energy is transferred from one organism to the next in a community.

- Food chains start with green plants (**producers** of food).
- Animals that feed on plants or other animals are **consumers**.
  - An animal that only eats plants is called a **herbivore**.
  - An animal that feeds on other animals is called a **carnivore**.

Plants and animals belong to more than one food chain. So we can join food chains to form a **food web**.

The diagram shows part of a seashore food web.

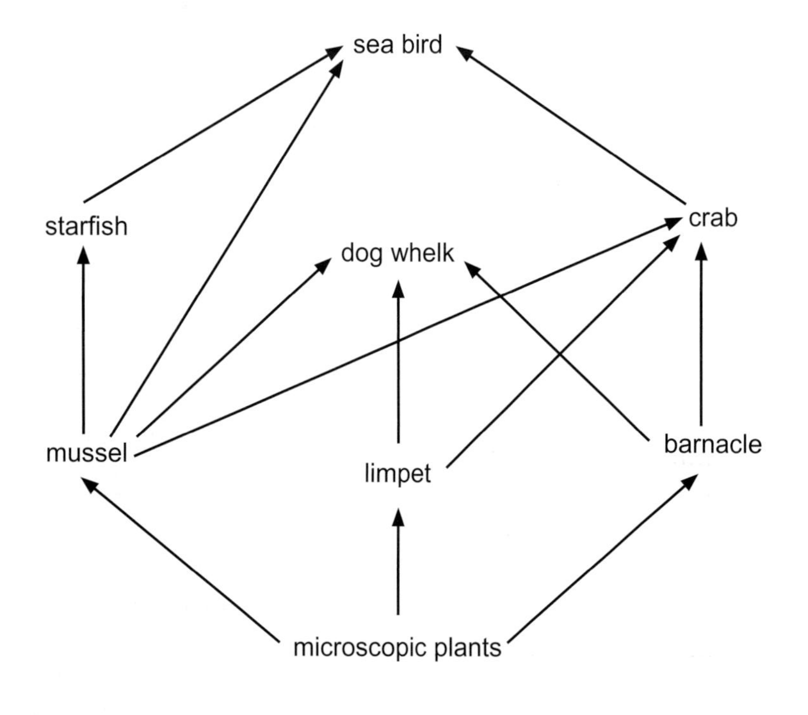

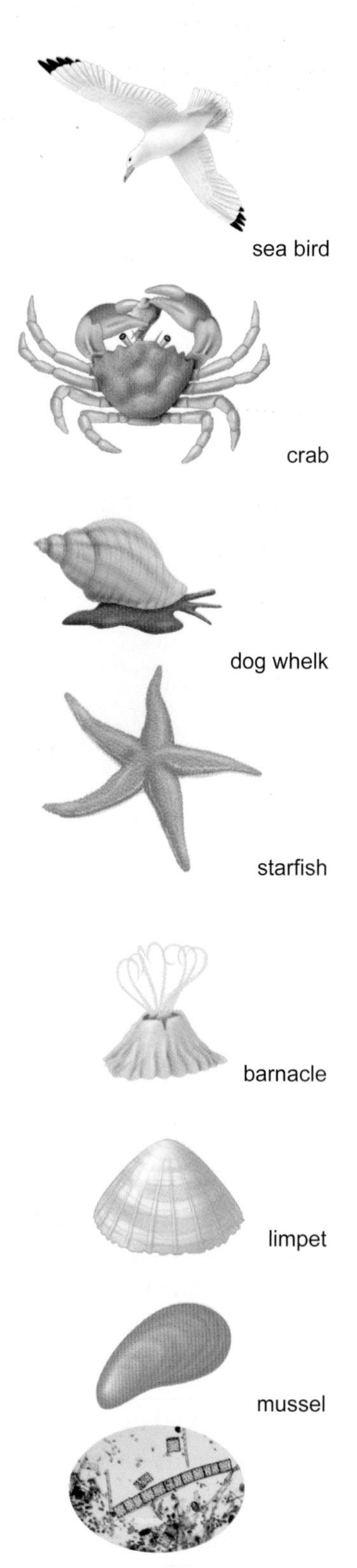

Question 1 2 3 4 5

Using food webs

We can use food webs to predict what will happen if the size of the population of a particular plant or animal changes.

If the population of one organism goes up or down, it affects the rest of the food web.

If a disease kills the mussels:

- the number of limpets might decrease – the dog whelks have fewer mussels to eat, so they eat more limpets
- the number of sea birds might decrease, because they have lost one of their food sources.

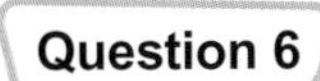

Pyramids of numbers

There is less energy for the organisms at each stage in a food chain. This is because living things use energy to move, grow and keep warm.

So, the number of animals gets smaller as you go along the food chain.

100 lettuces → 10 rabbits → 1 fox

We can show the change in population as we move along a food chain by a **pyramid of numbers**.

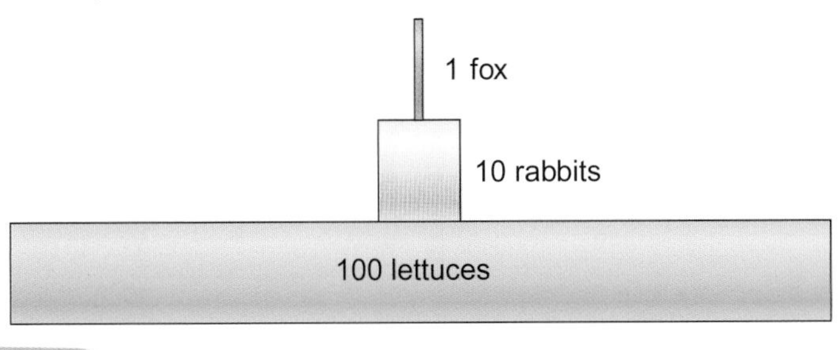

Question 8

A problem with pyramids of numbers is that they do not allow for the sizes of the organisms at each level of the food chain. So a pyramid of numbers for a tree is not a pyramid shape.

An oak tree is very big compared with a caterpillar. There is only one oak tree but it has thousands of leaves for caterpillars to feed on.

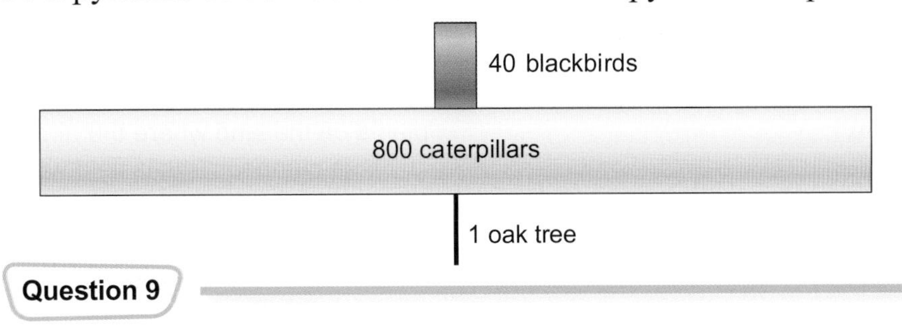

Question 9

Review your work

Summary ➡

8D.HSW Ways of gathering evidence

You should already know | Outcomes | Keywords

Types of investigation

Some ways of answering a question in science are:

- using secondary sources such as books or the Internet. Information collected by other people is called **secondary data**.
- doing field work, including surveys by sampling.

Question 1 | 2 | 3 |  4

Martin is researching the impact on the environment of a planned new airport. He is using scientific books and papers. We call this a desk study.

Information

Scientists publish reports of their research in **scientific papers** in scientific journals. Martin found papers on the Internet and downloaded them. These papers are secondary sources of information.

Question 5

Martin studied the environment.

Sally surveyed the opinions of people who

- lived near the planned airport
- ran businesses there.

Martin also uses traps to sample animals and **quadrats** to sample plants on the land where the airport is planned. This is field work.

Surveys of data about people, including people's **opinions**, often involve questionnaires.

When people give their opinions, they say what they think.

- Often, they have no evidence for their ideas.
- Sometimes, their opinions are **biased**. They are conclusions that, if correct, would benefit them in some way.

An example is the case of the Little Owl.

Investigating the Little Owl

In the 1890s, a few pairs of the Little Owl were introduced into Britain.

By the 1930s, they were

- widespread
- being accused of killing birds, including pheasant chicks.

Gamekeepers, who look after pheasants, killed Little Owls when they got the chance.

Little Owls are predators. In 1935, no one knew what they ate.

In 1935, the British Trust for Ornithology wanted evidence of what the Little Owl ate. A naturalist called Alice Hibbert-Ware had already watched Little Owls and dissected their pellets. So the Trust asked her to find out what they ate.

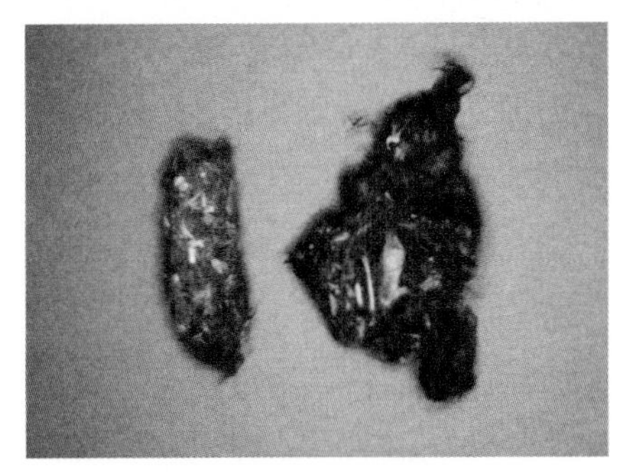

The Little Owl makes pellets containing fur, bones and other parts that it can't digest. It spits them out.

What Alice did	She wanted to know…	She found…
Experiment 1 Alice analysed pellets from captive owls.	… if what was in the pellets matched what the owls had eaten.	… that they matched.
Experiment 2 She analysed samples sent to her from dozens of places. These included pellets and samples from stomachs and nest holes.	… what the owls ate.	… mainly fur and bones of rodents, and the hard parts of insects and worms. The remains of birds were in the nest hole of only one owl.
She looked at reports from gamekeepers and landowners.	… what sort of evidence they had.	… gamekeepers reported the loss of game birds to owls, and landowners reported what their gamekeepers had told them.

Question 8

11

8D Questions

8D.1

1 What does the word habitat mean?

2 a Name the two main groups of animals.

b Which of the two groups has an inside skeleton?

3 a Write down the five groups of vertebrates.

b For each group, write down one feature that makes it different from the other groups.

4 Name five groups of invertebrates.

5 Name two types of vascular tissue in vascular plants.

6 Why do we divide plants into groups?

7 Mosses and liverworts often live in damp places.
Suggest why.

8 Vascular plants can live in more types of habitat than non-vascular plants.
Suggest why.

8D.2

1 Write down one example of innate behaviour and one example of learned behaviour.

2 Why is a good food supply important during the breeding season?

3 Sex hormones and day length affect breeding.
Which of these is an external and which is an internal factor?

4 Learning to build a safe nest benefits an animal.
Explain how.

5 Describe one other example of an external factor affecting breeding season.

6 Horseshoe crabs come out of the water to breed mainly during high tides at night.
Suggest two advantages of this timing.

7 Write down three things that might make a bird choose one mate rather than another.

8 Suggest how nesting in colonies (large groups) and making alarm calls help black-headed gulls survive.

continued

9 What innate feeding behaviour is the same for both of these gulls?

10 Young black-headed gulls run away from other gulls but young kittiwakes hide.
Suggest why they behave in different ways.

11 What do we call it when an animal recognises and follows another animal, usually its mother?

12 Explain why these ducklings are following Konrad Lorenz.

13 Suggest why imprinting is useful for survival of young.

8D.3

1 Write down <u>one</u> advantage and <u>one</u> problem of doing field work on a rocky shore.

2 Molly's friend David said, 'We can just count all the plants and animals that we see.'
Explain why this is not a good idea.

3 Write down <u>two</u> safety points that the class needs to think about before working on the rocky shore.

4 It was a good idea for several groups to do the same experiment.
Explain why.

5 Which of these areas is covered by the sea for the longest time?

6 Write down some patterns that you can see in Molly's table.

7 Molly's friend Patrick suggested that it would be a good idea to take some of the limpets back to the laboratory to look at in more detail.
Why must they not do this?

8 Lichens take a long time to dry out.
Why is this useful for an organism that lives on the upper shore?

9 Explain why seaweeds have strong holdfasts.

10 Mussels, limpets and barnacles fasten themselves to rocks in different ways.

a Find out how each attaches itself to the rock.

b How does attachment to rock help these animals to survive?

11 Barnacles and mussels keep their shells tightly shut until the tide comes in.
Suggest <u>two</u> reasons for this.

12 Which area of the shore contained the largest population of barnacles?

continued

13 Molly said, 'More barnacles live on the middle shore than the upper shore because the barnacles on the middle shore have more time to feed from the water.'
Do you agree or disagree with Molly's conclusion? Explain your answer.

14 Molly made the conclusion just from looking at her own results. If she is a good scientist, what should she do to make sure that the conclusion is correct?

15 The smaller temperature variation makes it easier for barnacles to survive on the middle shore than the upper shore.
Write down one other factor that makes it easier to survive on the middle shore.

8D.4

1 From the food web, write down:

a the producer;

b one consumer.

2 Use the food web to find two things that crabs eat.

3 Write down two food chains in the food web that end with a sea bird.

4 Name one herbivore in the food web.

5 Name one carnivore in the food web.

6 What will happen to the crab population if all the mussels are killed?

7 Limpets feed on microscopic plants on the rocks.
What effect will an oil spill on the rocks have on the food web?
Explain your answer.

8 Draw the shape of the pyramid of numbers for each of these food chains.

a dandelions → rabbits → fox

b microscopic plants → insect larvae → perch → pike

9 Draw a pyramid of numbers for each of these food chains.

a 100 lettuces → 10 000 slugs → 100 thrushes → 1 hawk

b 1 rose bush → 10 000 greenfly → 1000 ladybirds

8D.HSW

1 What is secondary data?

2 Where can you find secondary sources of information?

3 Look at the pictures. Martin is investigating the likely effect on the environment of a new airport in several ways.
Describe the kind of investigation that each picture shows.

4 Suggest two kinds of scientific report that Martin looks for.

5 Martin didn't just use his computer for finding information.
In your group, think of ways that Martin might use his computer in his investigation.

6 What did gamekeepers say about the Little Owl?

7 What is an owl pellet?

8 Little Owl was on trial. What sort of evidence was needed to prove whether it was guilty or innocent? Hint: look at the pictures.

9 In the passage about the Little Owl, find examples of:

  a hearsay – when one person reports what another person said. Courts don't accept hearsay. It is not evidence.

  b a biased opinion – when a particular idea is useful to someone.

  c valid evidence.

10 You could conclude that the information from gamekeepers and landowners was just biased opinions.
Why might gamekeepers want to believe that the Little Owl preyed on game birds?

11 What evidence from the pellets suggests that the Little Owl is not a pest, but is a useful bird?

8E.1 Materials

Scientific enquiry

You should already know

Outcomes

Keywords

Materials

Wood is the material that makes up this pepper mill. The pepper mill is an object. Wood is a natural material.

This pepper mill is made from wood. Wood is the **material** the mill is made from.

A material is not the same as an object. Think about a plastic ruler.

- The ruler is an object.
- Plastic is the material.

Some materials are made out of combinations of other materials.

Concrete is made from sand, cement and small stones.

But sand, cement and stones are also all made from other things.

Scientists have worked out that all materials are made from a few substances. These substances are called **elements**.

Question 1

What is an element?

There are about 100 elements that make millions of substances, just like letters make words.

Substances are made up of particles.

Some substances are made from only one type of particle. They are not combinations of other substances in any way. They are pure substances. They are called elements.

Elements combine to make other substances. This is like the letters in the alphabet. All the words in the English language are made from 26 letters.

There are about 100 elements. Any material that is not an element is made up from some combination of the elements.

Examples of elements are oxygen, hydrogen, copper and gold.

Materials like wood, water, ink, paper, calcium carbonate and hydrochloric acid are all made from combinations of elements.

For years people tried to find the elements that made gold. Unfortunately, gold is an element. So you cannot make it from something else!

Question 2  3

About elements

You can see that some things are not elements just by looking. If you look at concrete under a microscope, you can see it is made of pieces of sand and different coloured crystals. It cannot be an element if it is made from more than one thing.

Wood is another material that you can tell is not an element just by looking.

Concrete under a microscope.

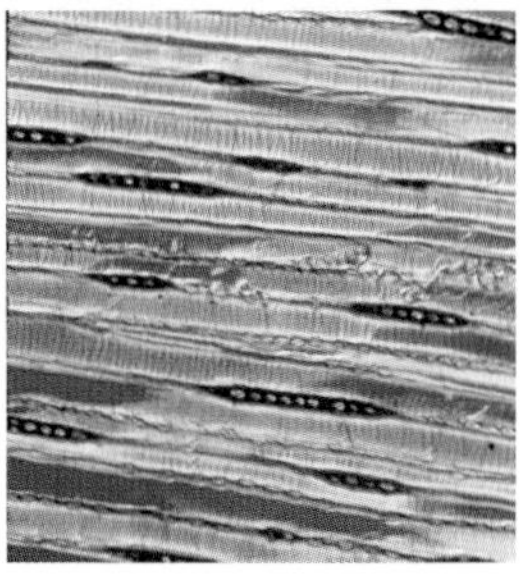

Wood under a microscope.

Water is not an element.

- If you pass an electric current through water, it splits up.
- The water splits up into hydrogen and oxygen.
- Water is made from the two elements hydrogen and oxygen.

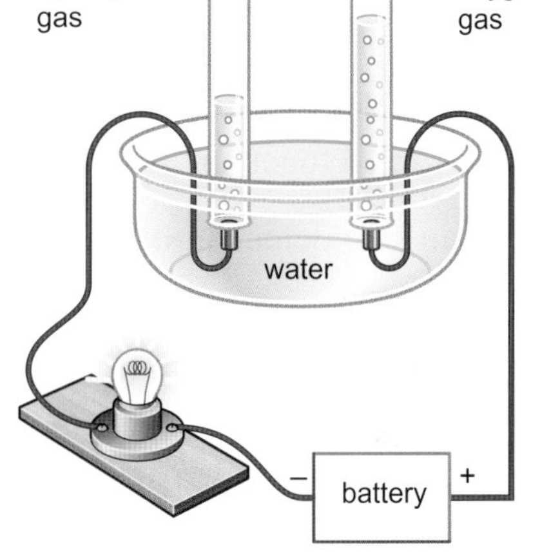

Electricity splits water into its elements.

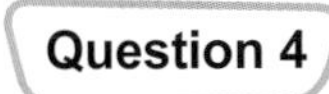

Question 4 **5**

Some elements have been known since ancient times, like gold and copper. Others, like radium, have only been discovered during the past hundred years or so.

The elements are not all as common as each other. The chart shows how common different elements are in the Earth's surface. If gold was as common as aluminium, it would probably not be worth as much!

There are about 100 elements. About 80 of them are metals.

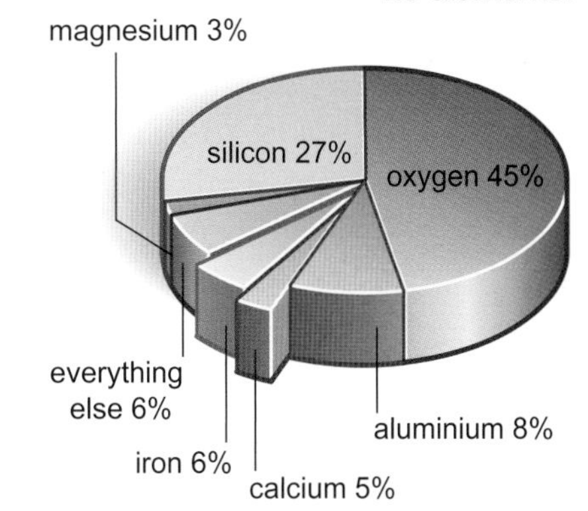

Pie chart showing how much of these elements there is in the Earth's crust.

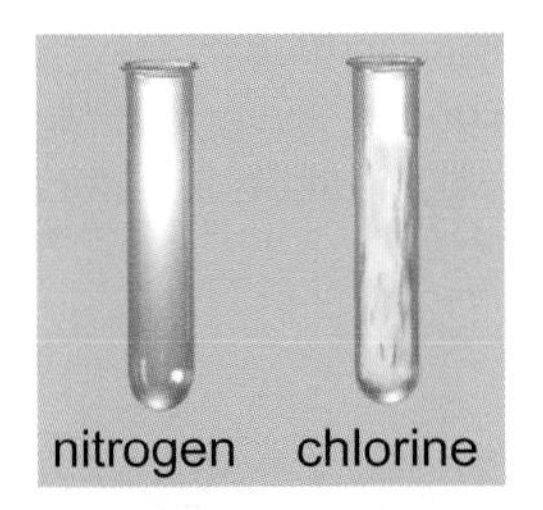

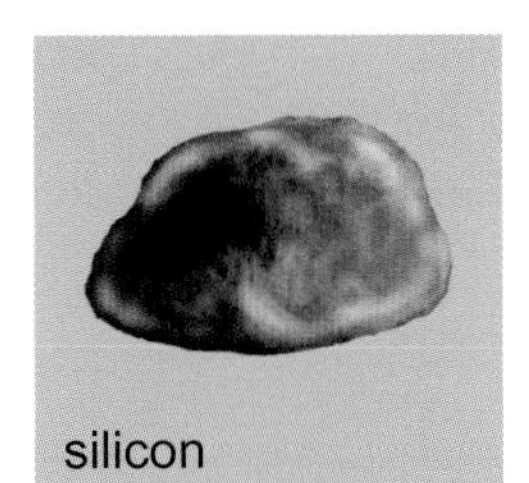

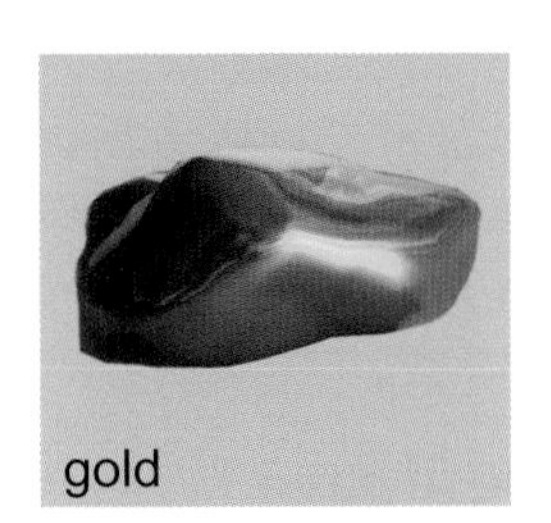

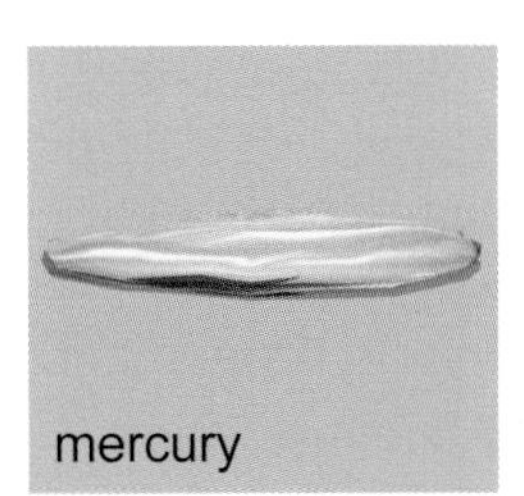

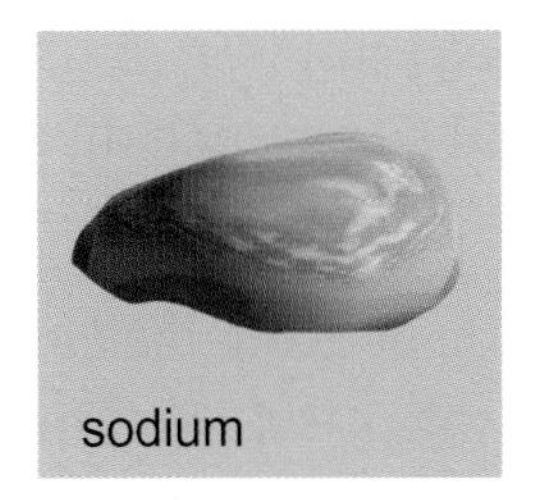

All of these are elements.

Question 6

8E.2 Atoms

You should already know | Outcomes | Keywords

What are atoms?

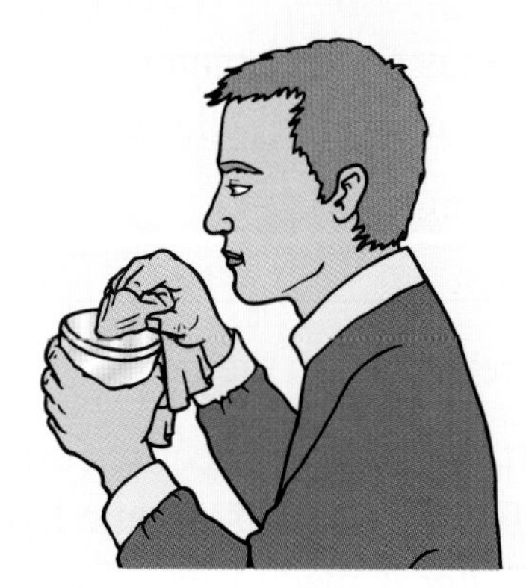

You can smell the aluminium. This is evidence for the existence of aluminium atoms.

Over 2000 years ago, a Greek philosopher called Democritus suggested that everything was made of small particles. He called them 'atoms'. Atoms are too small to see even with a microscope.

The **particles** that make up matter are single **atoms** or groups of atoms joined together.

You can also detect atoms even though you cannot see them, like this:

- Polish a piece of aluminium with a cloth.
- You will be able to smell the metal.
- Some aluminium particles have gone into the air and into your nose.

This coin from the time of Elizabeth the First is made from silver atoms.

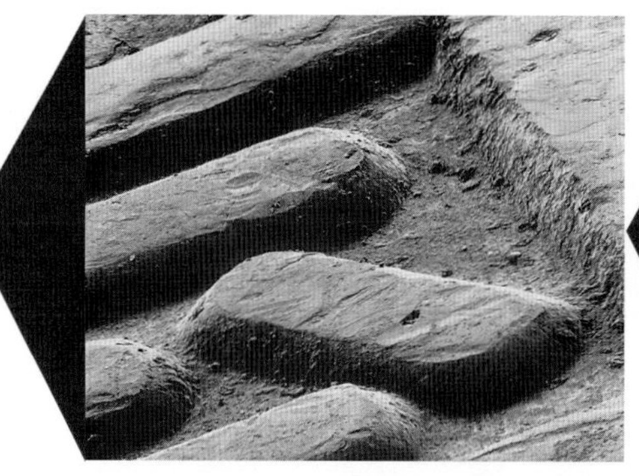

If you look at it under a microscope, the atoms are too small to see.

If we could see the silver atoms, they would look something like this.

Question 1 | 2

An element is all made from the same type of atom.

Gold only contains gold atoms, and silver only contains silver atoms. Gold and silver are both elements.

Water contains hydrogen and oxygen atoms. Water is not an element – it is made up from the elements hydrogen and oxygen. We say that water is a **compound** because it is made of more than one element joined together.

Substance	What is in it
carbon dioxide	carbon, oxygen
limestone	calcium, carbon, oxygen
salt	sodium, chlorine
iron	iron

Question 3 | 

Symbols

The atoms of different elements have their own **symbols**.

- All the symbols begin with a capital letter.
- Some symbols have a second letter in lower case.
- No symbol has three letters.

Name	Symbol	Metal or non-metal?	Solid, liquid or gas at 20 °C?	Colour	Year discovered
bromine	Br	non-metal	liquid	brown	1826
calcium	Ca	metal	solid	grey	1808
carbon	C	non-metal	solid	black	ancient
chlorine	Cl	non-metal	gas	green	1810
copper	Cu	metal	solid	pink	ancient
gold	Au	metal	solid	gold	ancient
helium	He	non-metal	gas	colourless	1868
hydrogen	H	non-metal	gas	colourless	1783
iron	Fe	metal	solid	grey	ancient
magnesium	Mg	metal	solid	grey	1808
mercury	Hg	metal	liquid	silver	ancient
nitrogen	N	non-metal	gas	colourless	1772
oxygen	O	non-metal	gas	colourless	1774
silver	Ag	metal	solid	silver	ancient
sulfur	S	non-metal	solid	yellow	ancient

Scientists use the symbol to represent one atom of an element.

'Cu' means 'one atom of copper'. 'Fe' means 'one atom of iron'.

Molecules

'**Molecule**' is the name for some atoms joined together.

Molecules can be very big. Some molecules contain hundreds of atoms.

The particles of carbon dioxide are molecules. Each molecule is made of one carbon atom and two oxygen atoms stuck together.

In the gas oxygen, the atoms go round in pairs. A molecule of oxygen is made from two oxygen atoms stuck together.

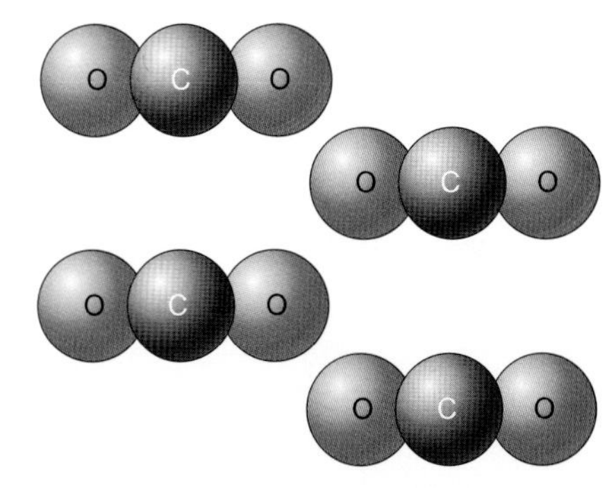

A model of carbon dioxide molecules.

Question 5  6

Check your progress

8E.3 Changes

You should already know | Outcomes | Keywords

There are two types of change – a **physical change** and a **chemical change**.

Physical change	Chemical change
No new substances are made. Examples: • boiling • melting • freezing • breaking into pieces	New substances are made. Examples: • burning • reacting with acid

Water looks different from ice but they are the same substance.

Other common examples of chemical changes are

- digesting food
- leaves rotting
- iron rusting
- cooking food.

Question 1 2

Using water.

Most substances are made from different atoms joined together. We call them **compounds**.

The table shows some examples.

Compound	Atoms in it	Some uses
water	hydrogen, oxygen	cooking, drinking, washing
common salt	sodium, chlorine	cooking, melting ice
carbon dioxide	carbon, oxygen	fizzy drinks, fire extinguishers

Using salt to help thaw ice on roads.

Question 3

New materials

We use chemical changes to make useful substances like iron.

We make iron from iron ore, which is found in the earth.

Iron ore contains the compound iron oxide. This is made from iron atoms and oxygen atoms combined.

A chemical change is needed to separate the iron from the oxygen. This is done in a blast furnace.

The diagram shows what happens.

One new material produced is iron. The blast furnace is so hot that liquid iron collects in the bottom of the furnace.

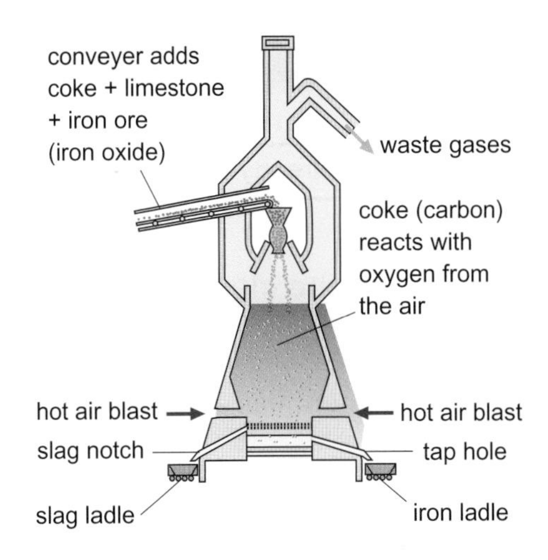

Making iron in a blast furnace.

Question 4

The properties of a substance are the things we can see and measure.

New materials made in a chemical change have different properties from the substances they are made from.

Burning magnesium is a good example of a chemical change.

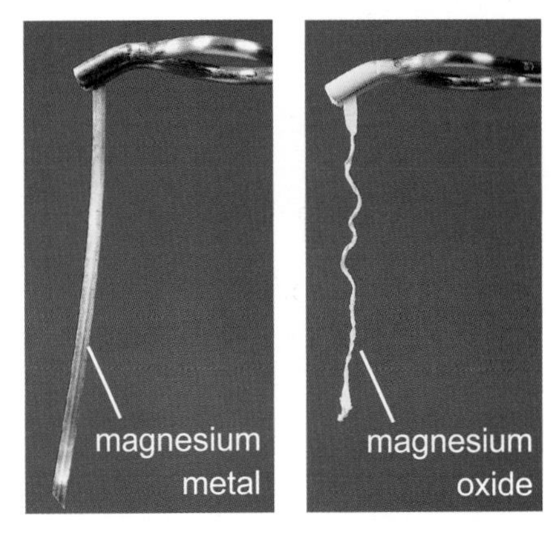

When the magnesium burns, it reacts with oxygen from the air.

Magnesium is an element. It has these properties:

- It is a soft, shiny metal at room temperature.
- It will conduct electricity.
- It can be bent and smoothed out into flexible strips.

Oxygen is an element. It has these properties:

- It is a colourless gas at room temperature.
- It has no smell.
- It does not conduct electricity.

When magnesium burns, the reaction is:

magnesium + oxygen ⟶ magnesium oxide

Magnesium oxide is a compound. It has these properties:

- It is a white, powdery solid at room temperature.
- It does not conduct electricity unless you melt it.

Magnesium oxide is nothing like magnesium or oxygen. It is a new substance.

Magnesium burns with a very bright, white flame.

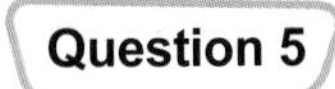

8E.4 Organising the elements

You should already know | Outcomes | Keywords

Patterns in the elements

Remember
Every element has a symbol made up from one or two letters.

There are about 100 different elements. Over the centuries, many different scientists have studied elements, trying to predict what they do in reactions.

After a lot of work by a lot of people, a very useful chart of the elements was put together. It is called the **periodic table**.

Group 1	2											3	4	Group 5	6	7	Group 0
					H hydrogen												He helium
Li lithium	Be beryllium											B boron	C carbon	N nitrogen	O oxygen	F fluorine	Ne neon
Na sodium	Mg magnesium											Al aluminium	Si silicon	P phosphorus	S sulfur	Cl chlorine	Ar argon
K potassium	Ca calcium	Sc scandium	Ti titanium	V vanadium	Cr chromium	Mn manganese	Fe iron	Co cobalt	Ni nickel	Cu copper	Zn zinc	Ga gallium	Ge germanium	As arsenic	Se selenium	Br bromine	Kr krypton
Rb rubidium	Sr strontium	Y yttrium	Zr zirconium	Nb niobium	Mo molybdenum	Tc technetium	Ru ruthenium	Rh rhodium	Pd palladium	Ag silver	Cd cadmium	In indium	Sn tin	Sb antimony	Te tellurium	I iodine	Xe xenon
Cs caesium	Ba barium	elements 57–71	Hf hafnium	Ta tantalum	W tungsten	Re rhenium	Os osmium	Ir iridium	Pt platinum	Au gold	Hg mercury	Tl thallium	Pb lead	Bi bismuth	Po polonium	At astatine	Rn radon
Fr francium	Ra radium	elements 89+															

Elements in the periodic table.

This periodic table shows the position of the elements in the table with their symbol and name.

The first thing to remember about the periodic table is that the columns with numbers are called **groups**. Elements in the same group have similar **properties**.

For example, lithium, sodium and potassium are in Group 1. They are all soft metals that react very dangerously with water.

a Lithium. **b** Sodium. **c** Potassium.

Question 1  2 3

Group 0

This group is on the extreme right of the periodic table.
It contains six elements. They are sometimes called the 'noble gases'. The table gives some information about the first three.

Name	Symbol	Information
helium	He	Colourless gas. Does not react with any other element. Used in aqualungs for divers.
neon	Ne	Colourless gas. Does not react with any other element. Used in electric advertising signs because it produces a red glow.
argon	Ar	Colourless gas. Does not react with any other element. Used in light bulbs so that the hot wire does not burn.

Question 4

Metals and non-metals

The chart shows a small part of the periodic table with illustrations of some elements.

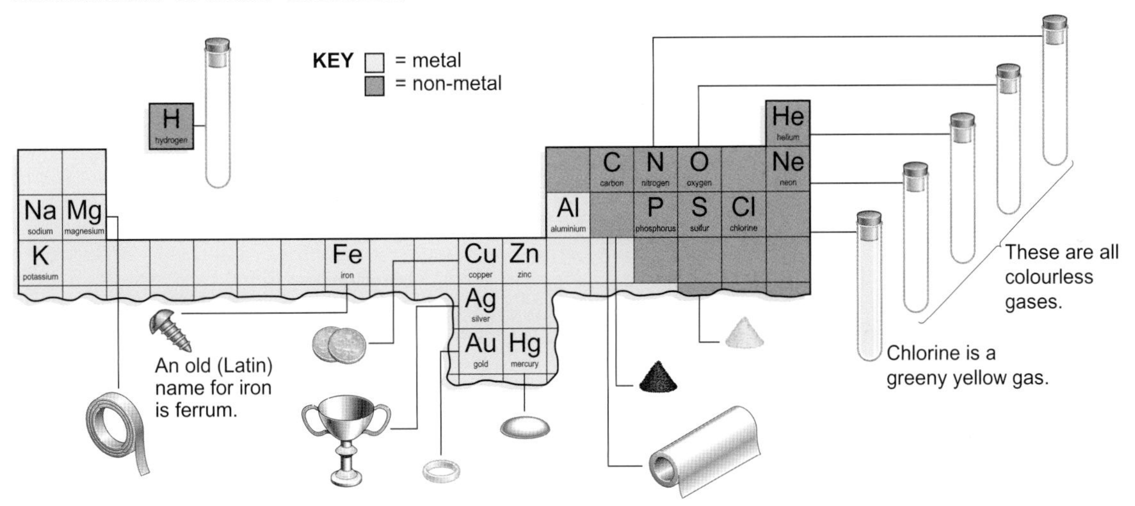

The elements in the part coloured blue are all metals.
The elements in the part coloured brown are non-metals.
You can carry the zigzag division down the full table to divide the metals from non-metals.

The study of the periodic table is a massive subject.
When scientists understand the information in it, they can make predictions about how elements will behave in reactions.
They can also make good guesses about the properties of elements they have never even seen!

Question 5  **6**

Review your work

Summary ➡

8E.HSW Developing the periodic table

You should already know | Outcomes | Keywords

Collaboration

The periodic table of the elements came from the work of many scientists. This happened over many years.

This is an example of people sharing work and ideas.

We call this **collaboration**.

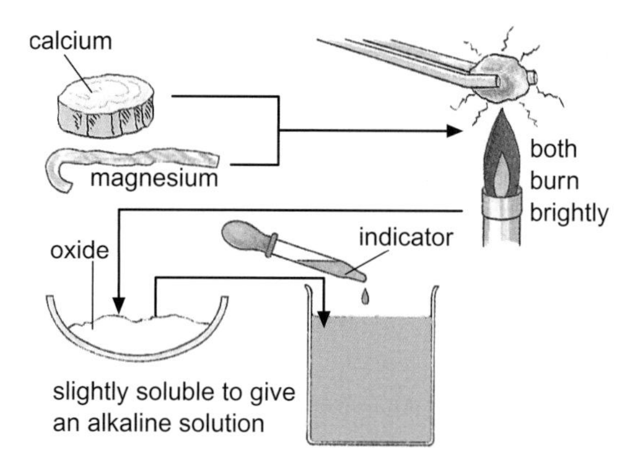

Calcium and magnesium behave in similar ways.

What scientists knew by 1860

By 1860 scientists already knew

- that everything is made from a few substance called elements
- that elements are made up from one type of atom
- that some elements behave in similar ways – for example, calcium and magnesium both burn brightly.

In 1860, the world's first chemistry conference took place in Germany. Scientists from all over the world attended it. This was an example of collaboration.

At the conference, the scientists published an up-to-date list of how heavy the atoms of different elements were.

Some scientists used the new information.

This led to the discovery of the periodic table of the elements.

Newlands' discovery

In 1864, a scientist called John Newlands wrote down a list of the elements he knew. He put them in the order of how heavy they were. He noticed that, if you counted seven along from any element, you came to an element with similar properties. If you write the elements out in rows, elements that behave in a similar way appear underneath each other.

Newlands' pattern had a problem. It only worked for the first 20 elements.

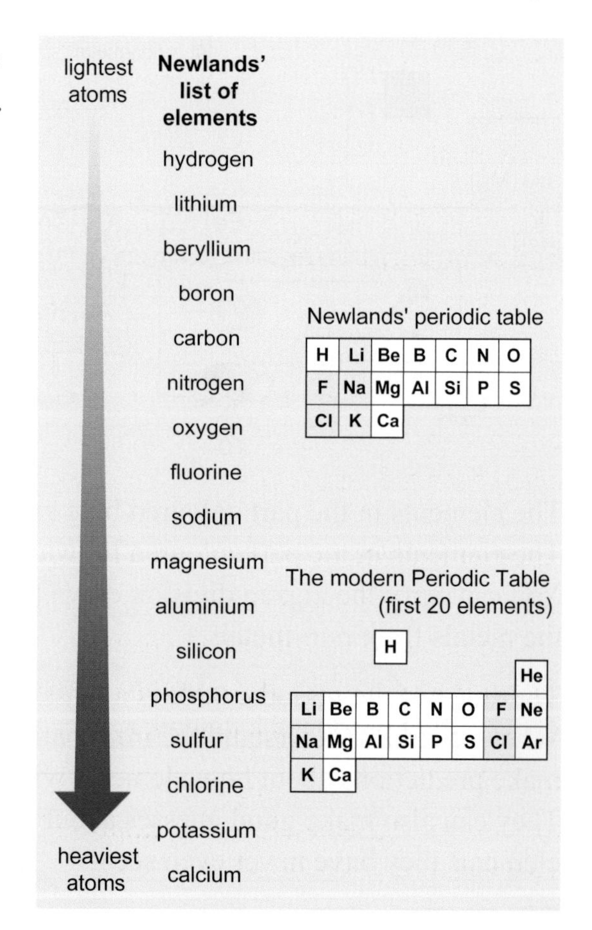

Question 1 2

Improving the idea

A Russian scientist called Dmitri Mendeleev improved the idea. He also used it to make **predictions** that could be tested.

Mendeleev had to solve some problems.

- At the time, only about 60 elements had been discovered.
- Some of the information about the atoms was wrong.

He solved these problems by

- leaving spaces where the properties of an element did not fit
- swapping elements round where he thought the data was wrong.

Mendeleev was **critical** of the evidence from experiments.

Mendeleev ended up with a table with some holes in it. Then he did a brave thing. He predicted what the missing elements would be like. He did this from what he knew of the other elements in the table. These predictions could be tested.

Three of the elements Mendeleev predicted were discovered within 15 years. Mendeleev was right!

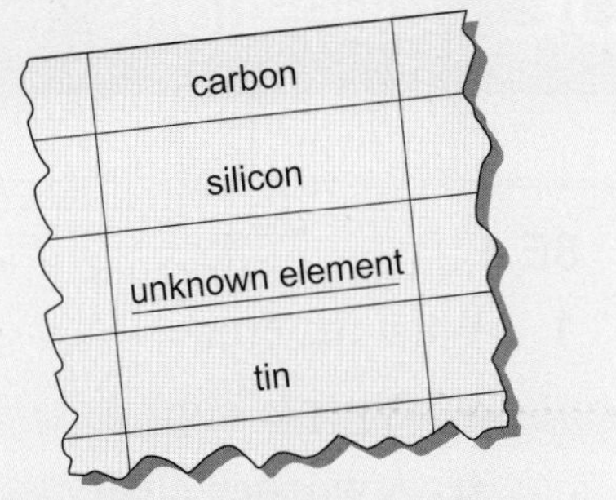

What Mendeleev said the unknown element would be like (in 1869)

- a grey metal
- its oxide would be white
- its chloride:
  would boil at less than 100˚C
  each cm^3 would have mass 1.9 g

The element germanium (discovered 27 years later)

- a grey metal
- a white oxide
- its chloride:
  boils at 86.5˚C
  each cm^3 has mass 1.8 g

The missing group

One thing that Mendeleev did not predict was a whole extra group of elements.

The noble gases, argon, neon, krypton and xenon were discovered by a Scot called William Ramsay. He worked out that these elements fitted in the periodic table as another group.

The discovery came from a **conflict** of evidence. The mass of nitrogen made from the air was slightly different from the mass of nitrogen made in other ways. Ramsay worked out that the nitrogen made from the air contained a small amount of argon.

Sir William Ramsay won a Nobel Prize for discovering the missing group in the periodic table.

Another problem

If you put the elements in order of how heavy they are, some of the elements do not fit into the pattern of the periodic table. Argon is an example. It fits before potassium but it is heavier. This conflict produced a new idea.

In 1913, a scientist called Henry Moseley worked out an order for the table based on how the atoms in elements were made up. This was a better idea and gives us the table we use today.

Group	1	2	3	4	5	6	7	0
			H					He
	Li	Be	B	C	N	O	F	Ne
	Na	Mg	Al	Si	P	S	Cl	Ar
	K	Ca						

relative atomic masses:
argon (Ar) = 40
potassium (K) = 39

Argon is heavier than potassium but it fits in the pattern before it.

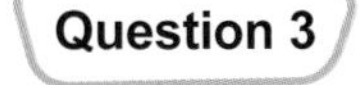

8E Questions

8E.1

1. What material is each of these items made from?
   - **a** A paper cup.
   - **b** A wooden chair.
   - **c** A steel blade.
   - **d** A cotton shirt.
2. What are elements? Give <u>two</u> examples.
3. Give <u>five</u> examples of substances that are made from combining elements.
4. List <u>three</u> everyday substances that you can tell are not elements just by looking.
5. Why is water not an element? What is it made from?
6. Find out who discovered radium and why radium is a very dangerous element.

8E.2

1. What are atoms?
2. Describe <u>one</u> way of detecting aluminium atoms even though you cannot see them.
3. What is the name for a substance made from more than one type of atom?
4. Copy out the table and add a third column that shows whether each substance is an element or a compound.
5. What are the symbols for these elements?
   - **a** Gold. **c** Iron. **e** Hydrogen. **g** Oxygen.
   - **b** Copper. **d** Lead. **f** Chlorine.
6. Draw an oxygen molecule. Use the diagram of carbon dioxide molecules as a guide.

8E.3

1. What is the difference between a physical change and a chemical change?
2. Give <u>two</u> examples of chemical changes.

continued

3 Give three examples of common compounds and list the elements that make them up.

4 Describe how iron is made from iron ore.

5 What reaction takes place when magnesium burns?

6 How do you know a chemical reaction has taken place when magnesium burns?
Give three examples of the properties of the product that tell you it is a new substance.

8E.4

1 What is the periodic table and what type of information does it show?

2 **a** What is the name for the columns with numbers at the top?
b What is special about the elements in the same column?

3 Suggest one reason why lithium, sodium and potassium all fit well in the same group.

4 Suggest one reason why a gas that does not react is useful in a light bulb.

5 Give one difference between carbon and aluminium that you can tell from their positions in the periodic table.
Give a reason for your answer.

6 Name one element that has similar properties to sodium.
Give a reason for your answer.

8E.HSW

1 The elements sodium, potassium and lithium are all in the same group in the periodic table.
Find out what properties they have in common.

2 Why was the chemistry conference in 1860 an example of collaboration between scientists?

3 **a** Find out the key facts about Mendeleev, including when and where he was born, and when and where he worked.
b Find out which element in the periodic table is named after him.

4 **a** What honour did William Ramsay receive for his work on the periodic table?
b What tragedy occurred to prevent Henry Moseley producing more important ideas.

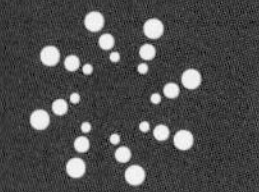

8F.1 Names and formulae

Scientific enquiry

You should already know | Outcomes | Keywords

What's in a name?

'Pass the sodium chloride, please.'

A compound has a **chemical name**, like carbon dioxide or sodium chloride.

Compounds sometimes have a name that we use in everyday language, like 'salt'.

The same compound can sometimes be found in forms that look quite different. It has the same chemical name but its common name is different. This can be for a range of reasons.

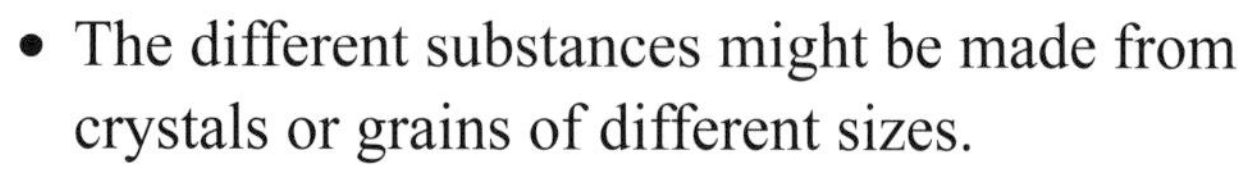

- The different substances might be made from crystals or grains of different sizes.
- The different substances might be mixed with different impurities.
- The way the atoms or molecules are arranged in the substance can make it look different even though the atoms inside the substances are the same. Graphite and diamond are both pure carbon but they look very different because of the way the carbon atoms are arranged inside.

Carbon (diamond).

When a substance melts or evaporates, it changes its appearance. Sometimes, its name changes as well. For example, ice, water and steam are all the same compound.

The table gives some more examples.

Common name	Chemical name	Elements inside it
salt	sodium chloride	sodium, chlorine
chalk	calcium carbonate	calcium, carbon, oxygen
limestone	calcium carbonate	calcium, carbon, oxygen
natural gas	methane	carbon, hydrogen
alcohol	ethanol	carbon, hydrogen, oxygen

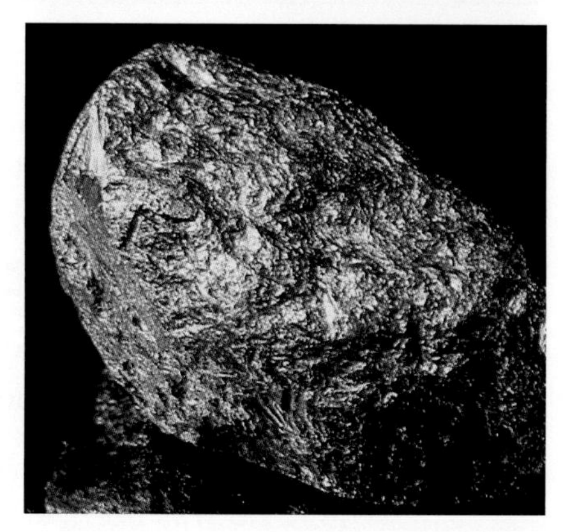
Carbon (graphite).

Question 1 2

What is a formula?

Scientist use a **formula** for each compound to show exactly what atoms are in each particle of the compound.

The formula is the same in any language. That is better than the name. The name might vary from one country to another.

- The symbol for an atom on its own means there is one of those atoms in a particle of the substance.
  Hydrochloric acid has the formula HCl. 'H' means one atom of hydrogen. 'Cl' means one atom of chlorine.
- If there are two or more atoms in the particle, then the number of them is written below the line after the symbol.Carbon dioxide has the formula CO_2. 'C' means one atom of carbon. 'O_2' means two atoms of oxygen.

Water is H_2O everywhere in the world.

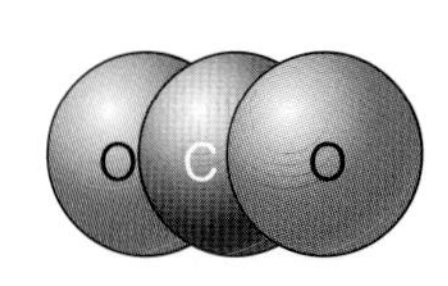

Carbon dioxide.

Here are some examples.

Chemical name	Formula	What it contains
calcium oxide	CaO	one calcium atom, one oxygen atom
magnesium sulfide	MgS	one magnesium atom, one sulfur atom
carbon dioxide	CO_2	one carbon atom, two oxygen atoms
calcium carbonate	$CaCO_3$	one calcium atom, one carbon atom, three oxygen atoms

In science, the plural of formula is usually **formulae**, not formulas.

The table above has a formula for each compound.
There are four formulae in the table.

Breaking things up

Scientists do experiments on compounds to work out what the formula for the compound is.

Water can be split into hydrogen and oxygen using electricity. When you do that, you get twice as much hydrogen as oxygen from the water.

The formula for water is H_2O.

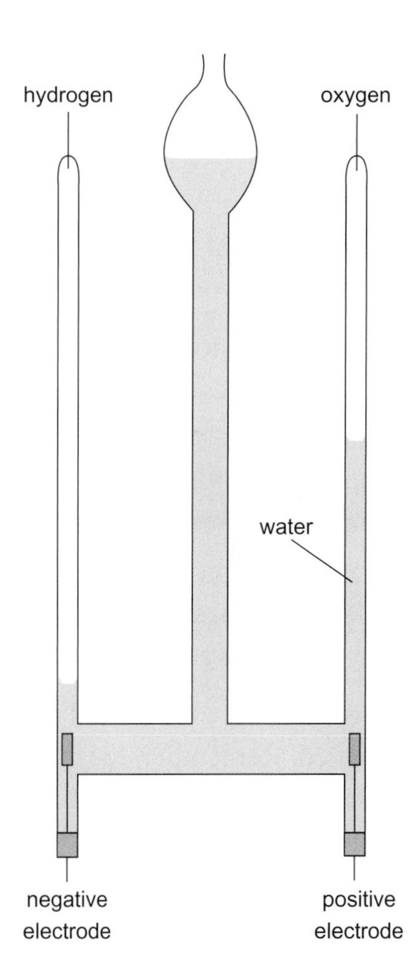

One way of splitting up water is by using electricity. This process is called electrolysis. A Hofmann voltameter is used for the electrolysis of water. Half as much oxygen is produced as hydrogen.

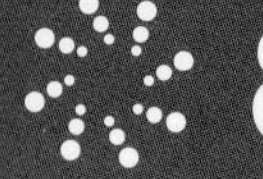

8F.2 Reactions and compounds

You should already know | Outcomes | Keywords

Products are different from reactants

Sodium is a reactive metal that will react violently with water to produce hydrogen.

Chlorine is a pale green gas that is very poisonous.

Sodium chloride is a compound made from the elements sodium and chlorine.

Sodium and chlorine react violently. They produce white crystals of sodium chloride. Sodium chloride is a new substance. It is a compound of sodium and chlorine.

Compounds are always different from the elements they are made from.

The product, salt, is safe enough to eat – provided you do not eat more than about 6 g a day. Chlorine on its own is a poison.

Sodium chloride is the salt that we add to our food to give it a salty taste.

Iron sulfide is a compound made from iron and sulfur. It looks different from sulfur and from iron. It is not magnetic like iron even though half its atoms are iron.

The word equation for the reaction between iron and sulfur is:

iron + sulfur ⟶ iron sulfide

Sometimes it can be quite difficult to tell if a new substance has been produced. For example, you do not always notice a gas being produced.

There are some clues to tell us if a chemical reaction has happened. Remember, these are only clues. Sometimes they can be misleading!

Here are the clues.

- A gas is produced.
- There is a colour change.
- Heat is produced.
- There is a change in mass.
- There is a change in appearance.

Question 1 | 2

Compounds and some of their reactions

Some compounds are very reactive. However, some compounds are less reactive. They will only react under certain conditions.

Scientists use terms to describe different types of reactions.

- Reactions with oxygen. Words used to describe reactions with oxygen include:
  - burning
  - **combustion**
  - oxidation
  - and even respiration.
- **Neutralisation** reactions. When an acid reacts with an alkali, the properties of the acid and alkali are cancelled out.
- **Precipitation** reactions. In some reactions between two solutions, one of the products is insoluble and it settles as a solid. The solid is called a precipitate. This happens when carbon dioxide makes lime water go cloudy.
- **Thermal decomposition** reactions. Some compounds break down or decompose when you heat them.

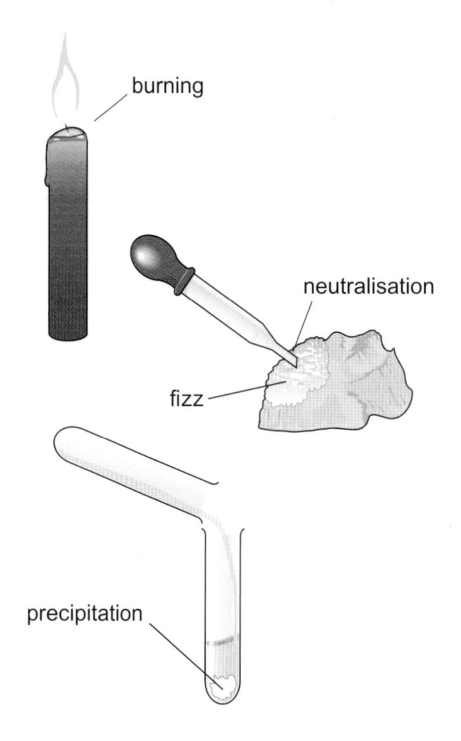

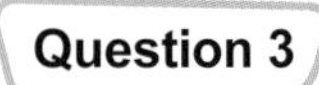

Limestone breaks down chemically when you heat it to about 550 °C. It undergoes thermal decomposition.

The chemical name for limestone is calcium carbonate.

The word equation for the reaction is:

calcium carbonate ⟶ calcium oxide + carbon dioxide

It is hard to tell that a reaction has happened. Calcium oxide looks like calcium carbonate. The carbon dioxide gas that escapes is invisible.

You can tell a reaction has happened by using water.

If you drip water onto calcium carbonate, nothing happens. If you drip water onto calcium oxide, you get a reaction that produces a lot of heat. The water turns to steam. It is a new substance.

The common name for calcium oxide is **quicklime**.

Calcium oxide is a very useful substance. People have been making it on a large scale in lime kilns since Roman times.

pieces of limestone (calcium carbonate)

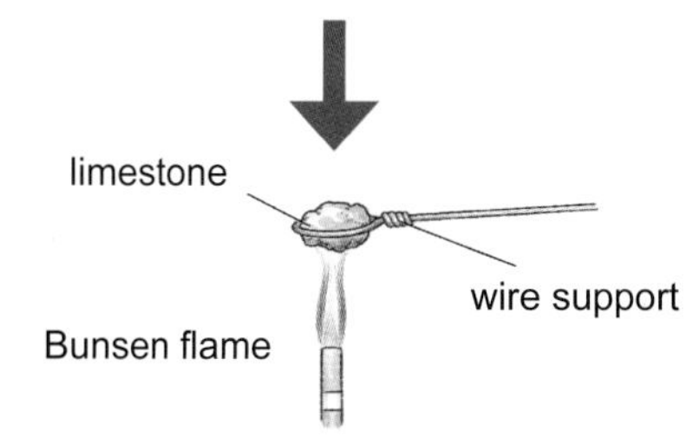

limestone changes into quicklime (calcium oxide)

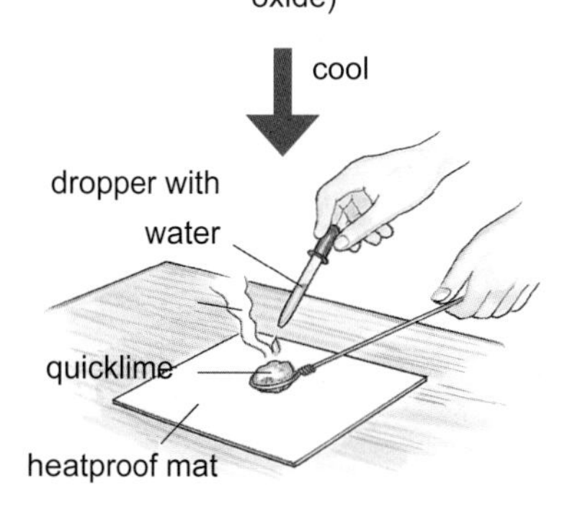

Question 4  

Check your progress

You should already know | Outcomes | Keywords

What is a mixture?

A pure element is made up of identical particles. All the atoms in it are the same.

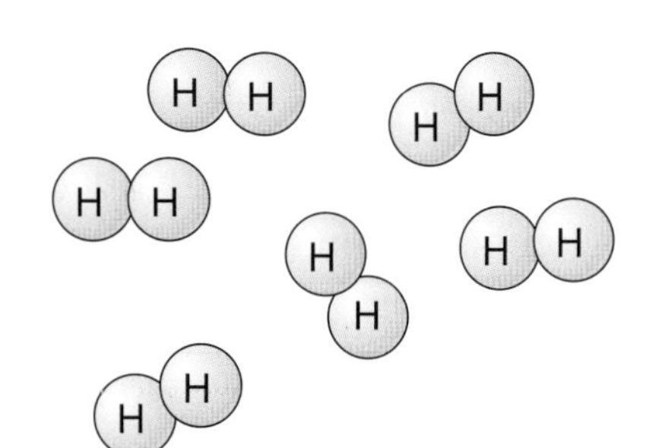

A particle diagram of a pure element.

A pure compound is made up of identical particles, but each particle contains more than one different type of atom joined together.

A **mixture** is not pure.

- A mixture is made of different particles just mixed together.
- Salt water, air and tea are all examples of mixtures.

Look at the particle diagrams.

The top diagram shows hydrogen molecules. It is not a mixture. There is only one substance present.

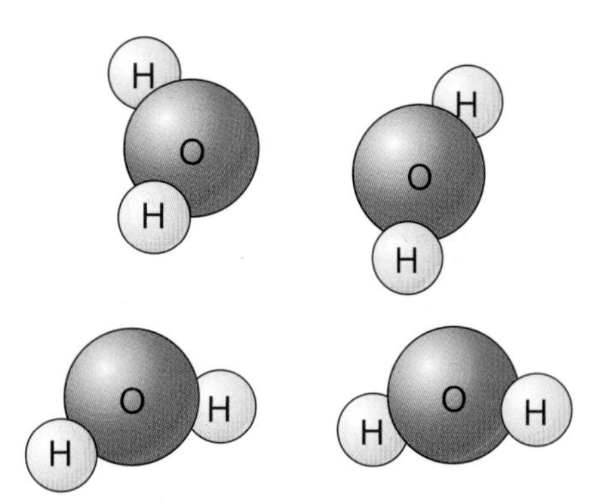

A particle diagram of a pure compound.

The middle diagram shows the compound water. It is not an element because it is made of hydrogen and oxygen atoms. It is not a mixture because there are only water molecules in it.

The bottom diagram shows a mixture of different gases.

The table gives the formulae and names for four of the gases in the diagram.

Formula	Description
O_2	two oxygen atoms joined make an oxygen molecule
H_2	two hydrogen atoms joined make a hydrogen molecule
NH_3	one nitrogen atom joined to three hydrogen atoms make a molecule of ammonia
N_2	two nitrogen atoms joined make a molecule of nitrogen

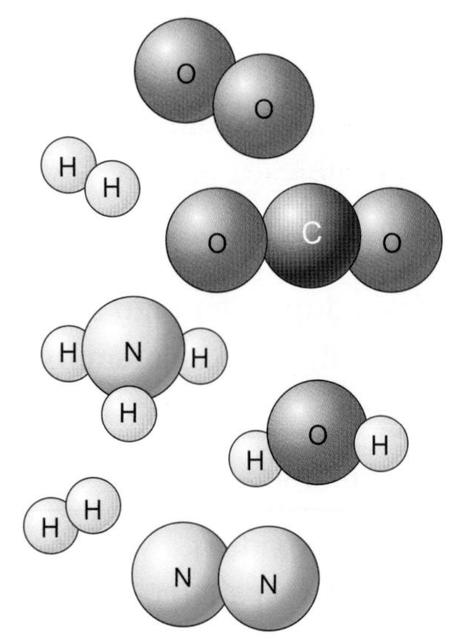

A particle diagram of a mixture.

Question 1 

Mineral water is a mixture

There are many examples of mixtures.

Mineral water is a mixture. Look at the label.

The different minerals listed in the water are dissolved in it.

The litre of water shown in the chart has almost half a gram of solids dissolved in it.

If that water evaporates or boils away in a pipe or kettle, the solids are left behind and they clog up the pipe or kettle.
This is called 'scale'.

When you dissolve something in water, you get a mixture.

Typical analysis mg/l	
Calcium	60
Magnesium	15
Sodium	46
Potassium	2.2
Carbonate ($CaCO_3$)	145
Chloride	155
Sulfate	1
Nitrate	5
Fluoride	0.1
Total dissolved solids	**453**

Question 3

Sea water is a mixture

Sea water is a mixture. It contains many different solids, called salts.

There is an average of 40 g of salts dissolved in every kilogram of water from the ocean.

The Dead Sea is not really a sea. The river Jordan flows into the Dead Sea but there is no outlet.

There are about 370 g of salts dissolved in every kilogram of water from the Dead Sea. The Dead Sea is the world's saltiest natural lake.

Over 43 billion tonnes of salts are dissolved in the Dead Sea.

The water going into the Dead Sea is not particularly salty.
However, the local climate is extremely hot and dry.
This makes the water evaporate and leaves the salts behind.
The Dead Sea is a bit like a huge evaporating dish.

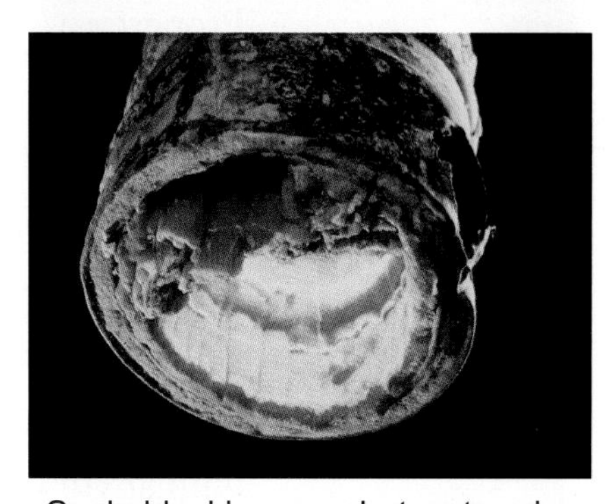

Scale blocking up a hot water pipe.

The water in the Dead Sea contains so much dissolved salt that you can easily float in it.

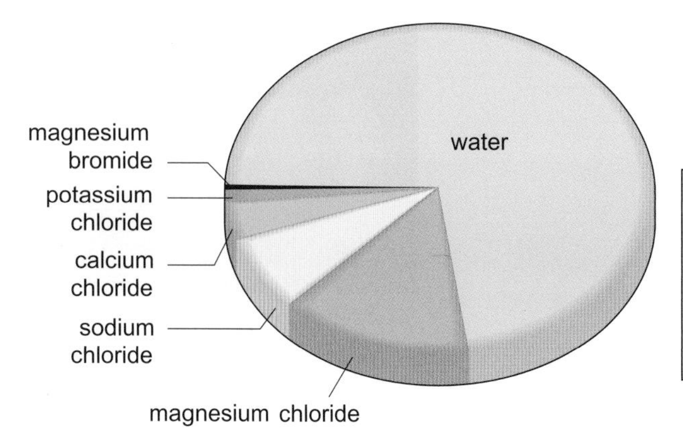

Mineral	% mass
magnesium chloride	14.5
sodium chloride	7.5
calcium chloride	3.8
potassium chloride	1.2
magnesium bromide	0.5
water	72.5

Mineral	Amount (billion tonnes)
magnesium chloride	22
sodium chloride	12
calcium chloride	6
potassium chloride	2
magnesium bromide	1

Salts in the Dead Sea.

Question 4 **5**

You should already know | Outcomes | Keywords

Air is a special mixture

The air is not a single substance.

The chart shows what is in the mixture we call air.

The table lists some important points about the gases in the air.

Gas	Percentage in air
oxygen	21
nitrogen	78
argon (and other noble gases)	1
carbon dioxide	0.035
water	6 – 0.1

nitrogen
water
carbon dioxide
argon
oxygen

The gases in the air.

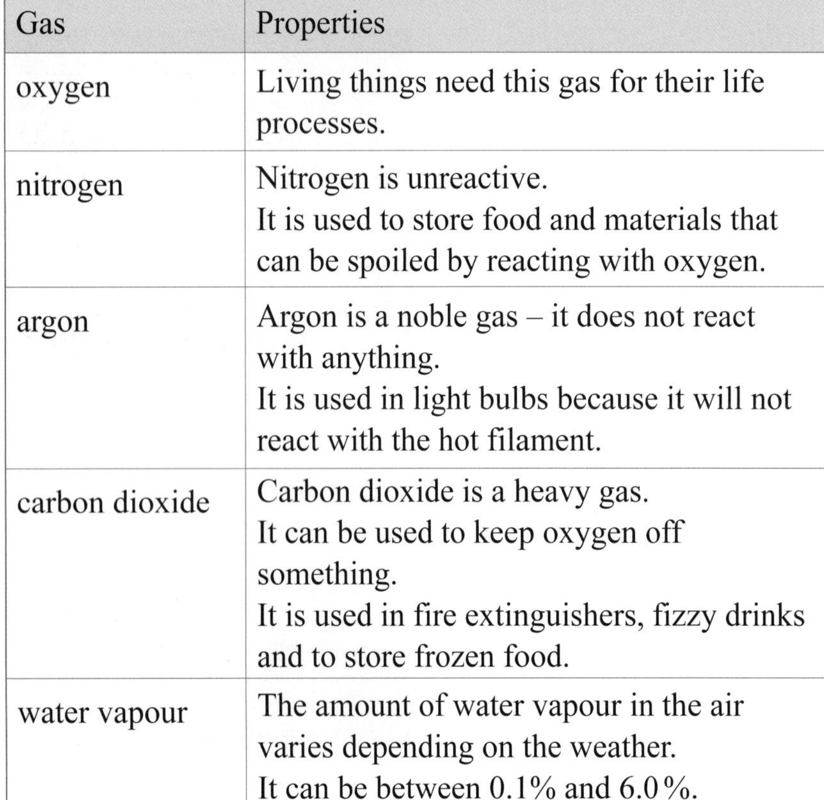

Gas	Properties
oxygen	Living things need this gas for their life processes.
nitrogen	Nitrogen is unreactive. It is used to store food and materials that can be spoiled by reacting with oxygen.
argon	Argon is a noble gas – it does not react with anything. It is used in light bulbs because it will not react with the hot filament.
carbon dioxide	Carbon dioxide is a heavy gas. It can be used to keep oxygen off something. It is used in fire extinguishers, fizzy drinks and to store frozen food.
water vapour	The amount of water vapour in the air varies depending on the weather. It can be between 0.1% and 6.0 %.

In hospital, patients with breathing or heart problems are sometimes given pure oxygen to breathe rather than air. This makes it easier for them to absorb oxygen into their blood.

Argon won't react with the metal filament in a light bulb, even when it is white hot.

Question 1  2 3

Separating the air

Because air is a mixture, it can be separated without using a chemical reaction just by cooling the air mixture down.

The chart shows the temperatures at which the gases change from a gas to either a liquid or a solid.

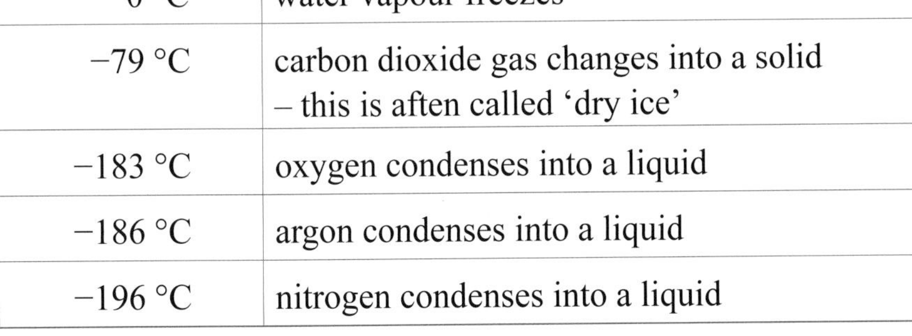

Temperature	What happens
0 °C	water vapour freezes
−79 °C	carbon dioxide gas changes into a solid – this is aften called 'dry ice'
−183 °C	oxygen condenses into a liquid
−186 °C	argon condenses into a liquid
−196 °C	nitrogen condenses into a liquid

Liquid nitrogen.

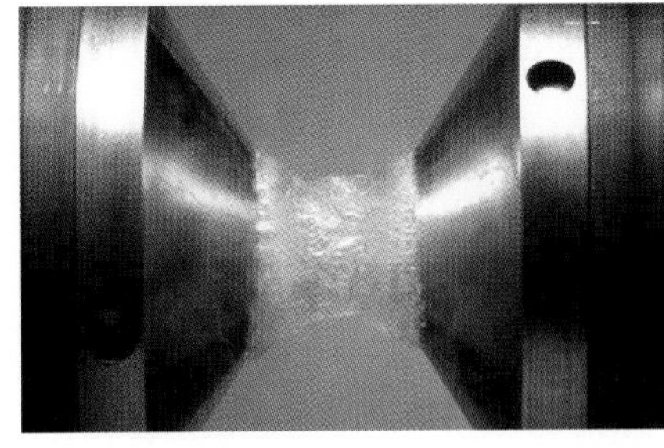

Liquid oxygen is magnetic. This is liquid oxygen held between the poles of a magnet.

Air is separated by **fractional distillation**. This is what happens.

- Air is cooled to below −196 °C.
- This changes all the gases in the air into liquids or solids.
- The mixture is warmed up slowly.
- As it reaches the temperatures shown in the table, each gas boils on its own.
- The gases are collected as they boil.

Nearly all of the oxygen used in medicine comes from liquefied air.

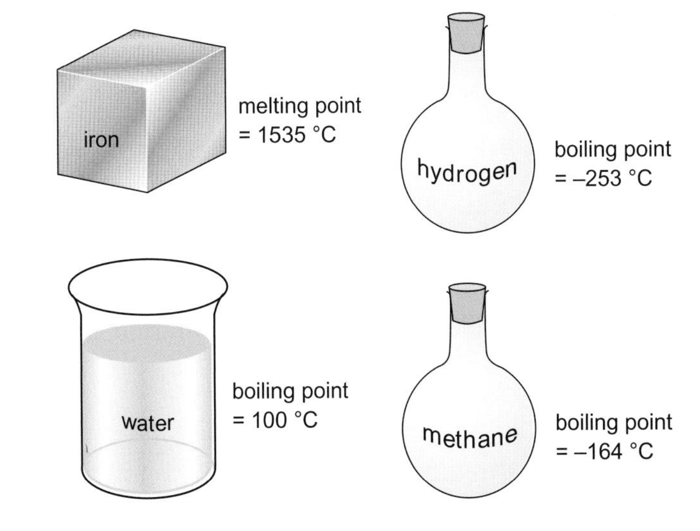

Melting points

Pure elements and pure compounds have exact boiling points and melting points.
The drawings and the chart below give some examples.

Substance	Boiling point	Melting point
sodium chloride	1413 °C	801 °C
oxygen	−183 °C	−214 °C
hydrogen chloride	−85 °C	−114 °C
mercury	357 °C	−39 °C

Mixtures do not have exact boiling or melting points.

You can tell that candle wax is a mixture by melting it and measuring the temperature when it melts. You find that it melts over a range of temperatures. It does not have an exact melting point. It is a mixture of substances.

Question 4  

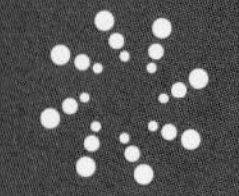

8F.5 Changing the atmosphere

You should already know

Outcomes

Keywords

In the beginning

You have seen what gases the **atmosphere** contains today. But when the atmosphere was first formed, about 4000 million years ago, it was very different.

- It was mainly **carbon dioxide** gas.
- It contained very little oxygen.
- There were small amounts of the gases **methane** and ammonia.
- There was some water vapour.

This is similar to the atmosphere on the planet Venus. Most of the atmosphere on Venus consists of carbon dioxide. Humans and other animals could not live on Venus because of the lack of oxygen.

Question 1 | 2

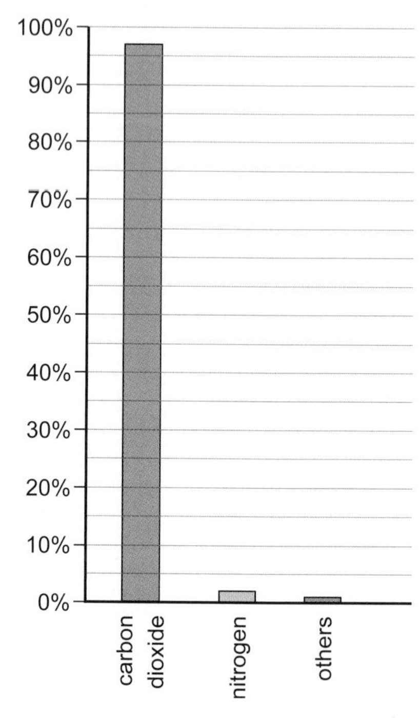

We need oxygen, but carbon dioxide poisons us. So we wouldn't be able to live on Venus.

Changing the atmosphere

About 3500 million years ago, there were very tiny plant-like organisms living in the sea. These tiny living things used carbon dioxide to make their food and gave off oxygen as a waste product. The levels of oxygen started to increase in the atmosphere.

Gradually, larger living things evolved. Eventually, there were plants growing in the oceans and on the land. The plants used carbon dioxide to make their food in the process we call **photosynthesis**.

carbon dioxide + water ⟶ food + oxygen

This increased the amount of oxygen and reduced the amount of carbon dioxide in the atmosphere.

By 2200 million years ago, oxygen levels were high enough to make iron rust. Banded red ironstone rocks are evidence of this.

Question 3 | 4

Locking up the carbon

Plants take in carbon dioxide and use the carbon in it to build material as they grow. When animals eat plants, the carbon passes from the plants into the animals. The carbon the plants take out of the atmosphere is locked up in the living things on the Earth.

Fossil fuels are made from the remains of living things. Other dead material like wood can also be used as fuel. When fuel is burned, carbon dioxide is released back into the atmosphere.

When living things die and rot, carbon is also released into the atmosphere as part of a gas called methane.

The carbon in sedimentary rocks like limestone originally came from the bodies of sea creatures that formed the rock.

This limestone is made from the remains of crinoids, or sea lilies. These animals are related to starfish.

Question 5

Human activity is changing the atmosphere

The level of carbon dioxide in the atmosphere has stayed about the same for a long time until just recently. Over the past 200 years, humans have been using fossil fuels. This has released carbon dioxide into the atmosphere. The levels are rising.

Other human activity such as changing limestone into lime also releases carbon dioxide into the atmosphere.

We began to use more and more fossil fuels during the industrial revolution. This began in the 1800s. The new machinery used coal, oil and gas.

Question 6

So what is the problem?

The levels of carbon dioxide are increasing. They are enough to cause the problem of **global warming**. Other gases we produce like methane also cause global warming. We refer to these gases as **greenhouse gases**.

Greenhouse gases produce a blanket around the Earth that keeps in the heat from the Sun, a bit like the glass in a greenhouse.

Humans need to live without causing global warming. We also need to do this without using up resources that cannot be replaced. This is called **sustainable development**.

Some scientists think that, unless we work very hard at this, the atmosphere will change enough to cause big problems. We may not be able to grow enough food, and sea levels may rise and swamp cities.

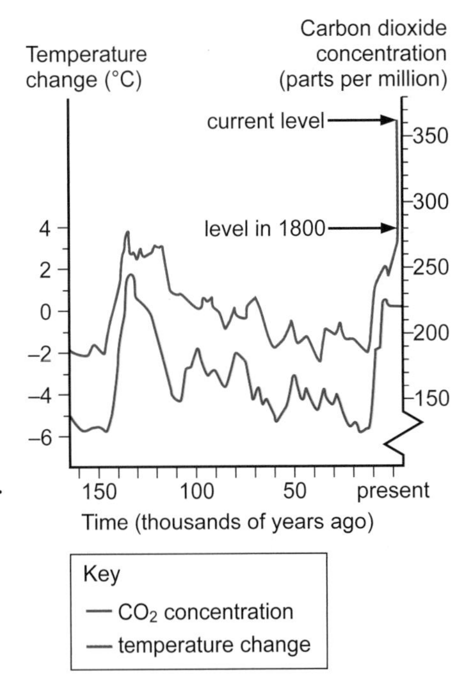

There is a link between the temperature of the Earth and the concentration of carbon dioxide in the atmosphere.

Question 7

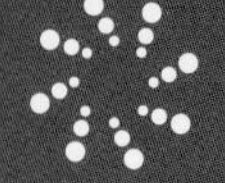

8F.6 Sustainable development

You should already know | **Outcomes** | **Keywords**

Human development

While humans have been on the Earth they have:

- improved their understanding of how things work;
- discovered what substances things are made from;
- discovered how to make new substances from the raw materials on the Earth.

They have also developed technology to make life easier.

Improvements in medicine and food production means there are many more humans on the planet than there were.

However, humans are using up the raw materials on the Earth to improve their standard of living. We use many different raw materials.

- We use ores to make metals.
- We use oil to make plastics.
- We use oil coal and gas for energy.

We also pollute the environment with waste materials.

Human development causes problems.

- We use raw materials to make new compounds, but these will eventually run out.
- We produce compounds like sulfur dioxide that damage the environment.
- We burn fossil fuels, which produces carbon dioxide and causes global warming.

In the past 200 years, the number of people on the planet has increased by about nine times.

Raw materials from the ground. (These will eventually run out.)

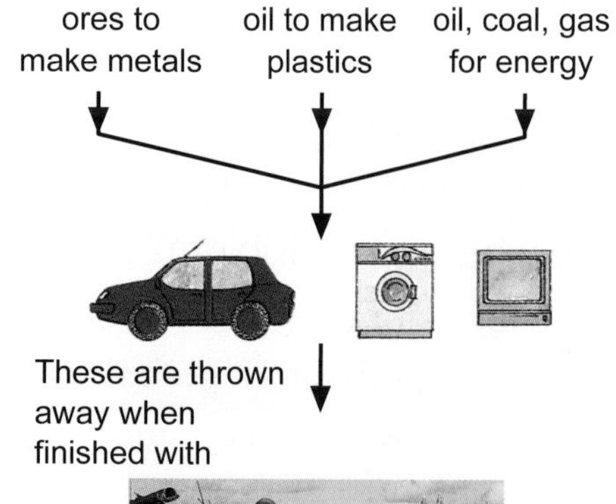

Question 1 **2**

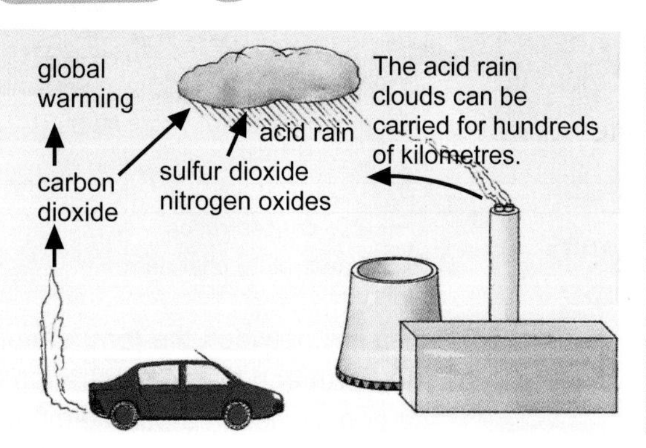

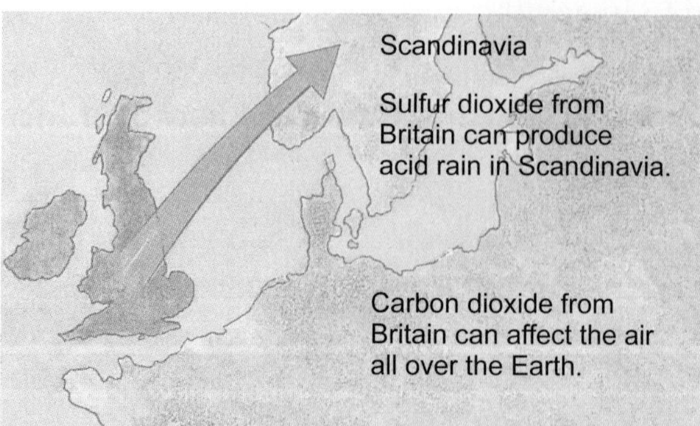

Needs and wants

Part of the problem is the difference between what people need and what people want.

Human beings have certain basic needs:

- food and water
- shelter
- clothing
- medicine
- energy sources for cooking and keeping warm.

In North America and Europe, we have what we need of these things and much more besides. This is not the same everywhere.

Many humans are very greedy. They may like a car that uses a lot of fuel, unnecessary air travel, luxury goods and a standard of living much higher than they actually need. There is a big difference between 'needing a new pair of trainers' and 'needing some food'.

Question 3

What is sustainable development?

There are more and more people on the Earth. We need to develop the ways we produce food and provide water, shelter and energy to meet the growing demand. We need to do this in a way that does not damage the environment.

Developing without damaging the environment or using up resources that cannot be replaced is called **sustainable development**.

The types of thing that governments are involved with to support sustainable development include:

- making people more aware of recycling materials
- setting out policies on the environment
- drawing up international agreements to limit the levels of carbon dioxide produced
- publicity to make people more aware of how their lifestyle affects the environment
- increasing tax on certain types of cars and on fuel

Question 4 **5**

The United Nations Environment Programme attempts to set standards and bring about co-operation between countries on environmental issues such as biodiversity, conservation of resources and climate change.

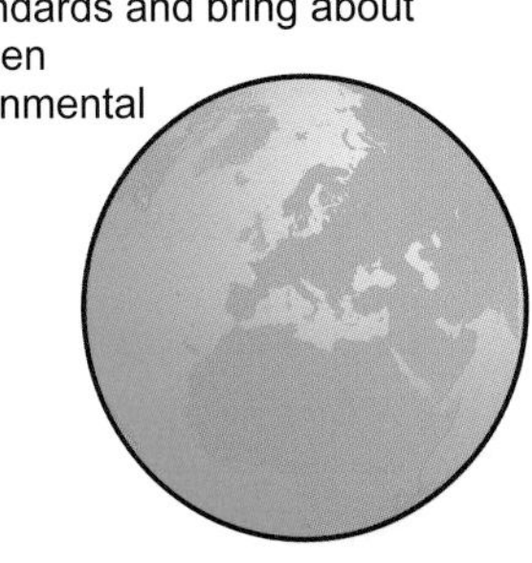

The UK government has policies on environmental issues such as biodiversity.

Review your work

Summary ➡

8F.HSW Extracting metals from the Earth

You should already know | Outcomes | Keywords

Metals

Metals are very important materials.

Most of the metals in the Earth are found in compounds. These compounds that contain metals are often called ores.

An important application of science is getting metals out of their ores.It looks simple in the laboratory.

The diagram shows a method of extracting a metal from an ore that contains a metal oxide.

- You heat the powdered ore with carbon powder.
- The carbon and the ore react and make carbon dioxide gas.
- This leaves the metal behind.

The word equation for the reaction is

metal ore + carbon ⟶ carbon dioxide + metal

Haematite contains iron. This ore is a type of iron oxide (iron combined with oxygen).

Malachite is copper carbonate (copper combined with carbon and oxygen).

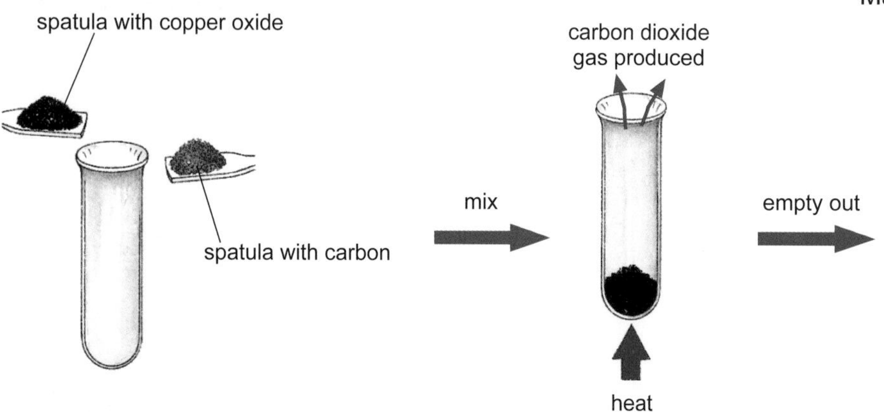

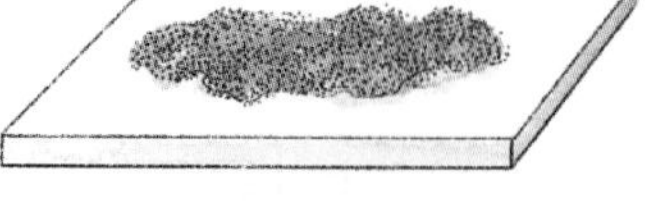

There are **ethical** and **moral** problems when you do this on a large scale.

- Ores have to be dug out of the ground.
- Ores contain a lot of waste material.
- The ore has to be transported to the extraction plant.

These facts have consequences for the environment.

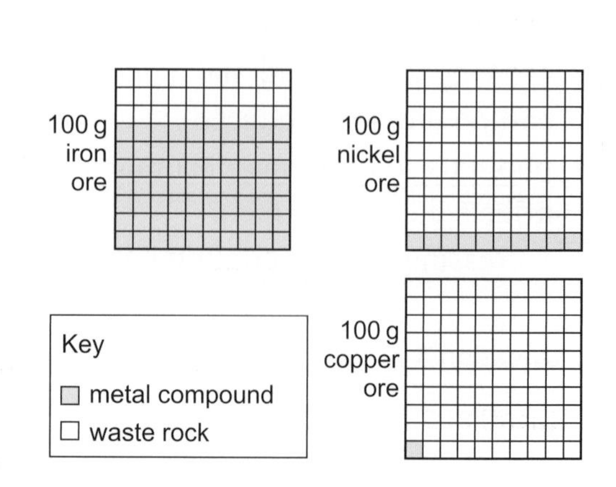

Question 1 | 2

Iron ore mine in Australia. Mining metals and metal ores can make huge holes in the ground.

Gold mine in Brazil. Quarrying and digging the metal ore produce a lot of dust, which pollutes the air.

Potash mine in Germany. Huge heaps of waste rock may be left behind. Wastes still contain metal compounds. These can pollute streams and harm living things.

The blast furnace

The blast furnace is used to extract iron from iron ore. It has several problems.

- A lot of energy is required to run it.
- It produces atmospheric pollution.
- It produces visual pollution.
- It requires good road or rail links to supply raw materials.
- It produces a lot of waste.

We pour carbon and iron oxide into the blast furnace. The iron oxide turns into iron metal. The reaction also makes carbon dioxide gas.

Question 3 **4**

Recycling metals

We can help to solve some of the problems associated with the production of metals by recycling the metals we use.

Aluminium is a metal we use a lot. It is extracted from its ore using electricity. If we recycle aluminium, we save energy.

This means that we produce less pollution from power stations. We have less impact on **global warming**.

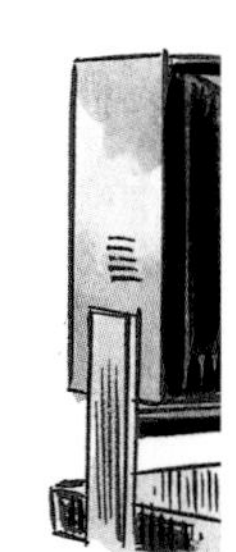

Did you know that recycling just one aluminium drinks can save enough energy to run a TV for 2 hours?

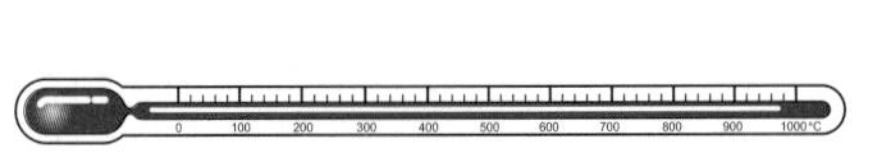

We have to melt the aluminium ore by heating it to a high temperature.

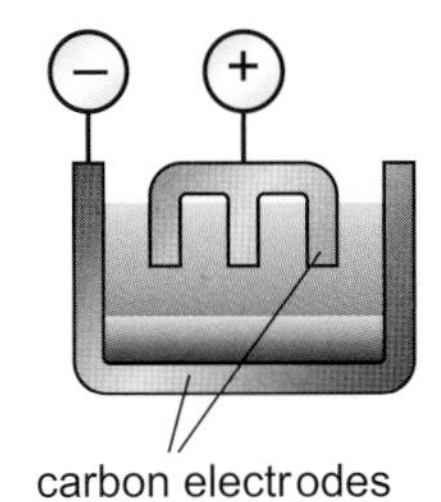

We extract aluminium from the molten ore using electricity.

Of course, recycling aluminium also uses energy, but only about a twentieth.

Question 5

8F.1

1 a What do graphite and diamond have in common?

  b Find out as many differences as you can between graphite and diamond, and draw up a table comparing them.

2 a What do chalk and limestone have in common?

  b Find out how they are different.

3 Give one example of the formula for a compound.

4 What is the difference between the meaning of the words formula and formulae?

5 a What is the formula for water?

  b Explain how we know that it is the formula for water.

8F.2

1 What happens when you put sodium in chlorine?
Give two ways in which the substance produced is different from sodium and chlorine.

2 List five clues that point to a chemical reaction taking place.

3 Give four different types of reaction.

4 Describe one way of showing that calcium oxide is a different substance from calcium carbonate.

5 a What is the name for the type of reaction in which calcium carbonate is changed into calcium oxide?

  b Explain why you might easily think that no reaction has taken place when you heat calcium carbonate strongly.

6 Find out why quicklime is a useful substance.

8F.3

1. What is meant by the term <u>mixture</u>?
   Give <u>three</u> examples.
2. Copy the table and add an extra column.
   Draw diagrams of the particles of each gas in the correct place in the third column.
3. What is the difference between the way a supermarket label uses the word <u>pure</u> and the way it is used in science?
4. How do we know that sea water is a mixture?
5. **a** What happens to a mixture of water and dissolved salts when the water evaporates?
   b Find out how people collect sea salt for cooking with.

8F.4

1. Which <u>two</u> gases make up most of the air?
2. Which gas in the air is essential for living things?
3. What properties of carbon dioxide make it useful in fighting fire?
4. If you cooled a sample of air down to −100 °C, what state would each different part of the mixture be in?
   Give reasons for your answers.
5. **a** What is meant by the term <u>fractional distillation</u>?
   b How is it used to get oxygen from the air?
   c Find out <u>one</u> other important use of fractional distillation.
6. Describe <u>one</u> way of telling if a substance is pure or a mixture.

8F.5

1. How have the levels of oxygen and carbon dioxide in the atmosphere changed from 4000 million years ago to today?
2. Why could humans not have lived on the Earth 4000 million years ago?
3. What caused oxygen levels in the atmosphere to increase?
4. Write down the word equation for photosynthesis.
5. How does the carbon from the carbon dioxide in the atmosphere end up as part of the bodies of living things?
6. What human activity has caused a rise in the level of carbon dioxide gas in the atmosphere in the past 200 years?
7. What is the problem with increasing the level of carbon dioxide gas in the atmosphere?

8F.6

1. What developments have led to there being more people on the planet in 2000 than there were in 1800?
2. Give three problems that are caused by human development.
3. List the basic needs for human beings.
4. What is meant by sustainable development?
5. What types of thing are governments doing to encourage sustainable development?

8F.HSW

1. A metal ore contains a lot of waste material.
   Why might this produce a problem at the extraction plant?
2. What sort of problems will the transport of a lot of material to and from an extraction plant cause for people living nearby?
   (Hint: think about how it might affect their health, the structure of their houses, the value of their homes if they own them. Will noise be a problem and what might they do about it if it is?)
3. List some of the problems for people living near a blast furnace.
   (Hint: a blast furnace produces a lot of fumes. There will be a lot of noise and light. The furnace will probably be operating 24 hours a day. Supplies will have to be brought in, and waste and products removed.)
4. Write a draft of a letter that you might send to your MP objecting to a blast furnace being built near to where you live.
5. Why is it a good idea to recycle aluminium cans?

8G.1 Different rocks

Scientific enquiry

You should already know

Outcomes

Keywords

Different rocks, different properties

Scientists who study rocks are called **geologists**.

There are many different types of rock.
Some rocks are very hard. Some are soft and crumbly.

Rocks occur in all sorts of shapes and sizes. They come in a wide range of colours.

Limestone is a fairly hard rock made from calcium carbonate.

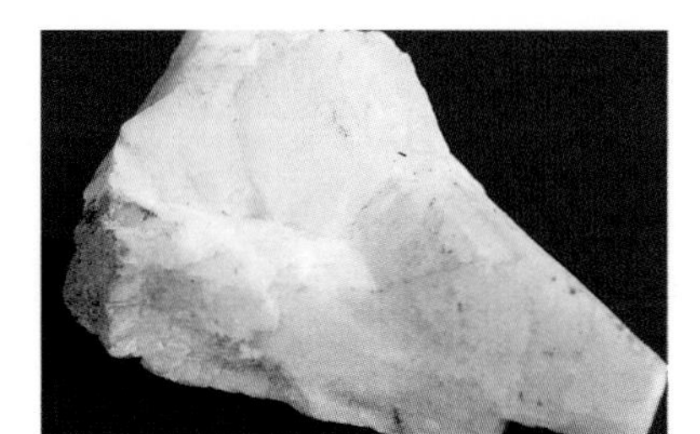

Marble is a rock made from calcium carbonate crystals. You can smooth and polish it.

Even though these rocks look different, they all contain the compound calcium carbonate.

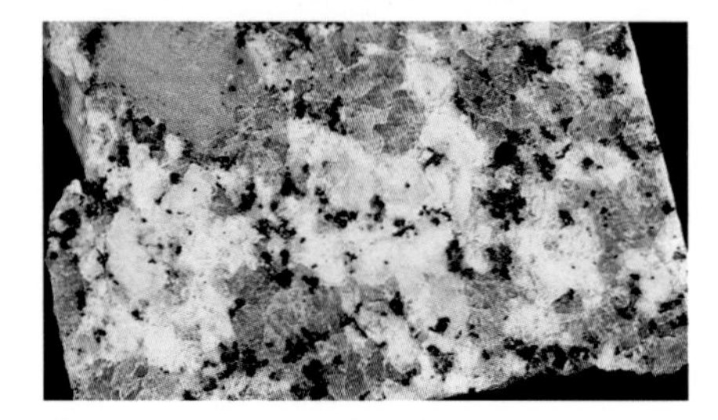

Granite is a very hard rock.

Sandstone is a soft rock. You can scratch it with your nails.

Slate is a hard rock made of layers. It splits easily into sheets.

Most rocks are made of a mixture of grains of different sizes.

Things like 'soft', 'crumbly', 'grain size' and 'colour' are called properties.

Different types of rock have different properties.

Some rocks are made from the compound calcium carbonate. But even though they are made from the same compound, they have different properties.

Because the rocks have different properties, they have different names.

Question 1 2

Rocks like sponges

Some rocks have very tiny gaps between the grains that make them up.

The gaps soak up liquids. They let liquids and gases pass through the rock like water passing through a sponge.

Rocks that let water, other liquids and gases through are called **porous** rocks. Sandstone is an example of a porous rock.

Some rocks have grains that fit very closely together with no spaces between them. These rocks do not let water soak into them. We say that they are non-porous.

Chalk is a porous rock.

Question 3

Porous and non-porous rocks combine to trap oil in the Earth's surface. Geologists can work out where oil might be found.

People can drill into the Earth to collect this valuable resource.

Oil forms in tiny drops in porous rocks. Because the rocks are porous, the water and oil can move up through the rocks. The oil floats on top of the water.

If the Earth's rocks are curved in a dome shape, with a layer of non-porous rock on top of porous rock, you can get oil trapped in the top of the porous rock.

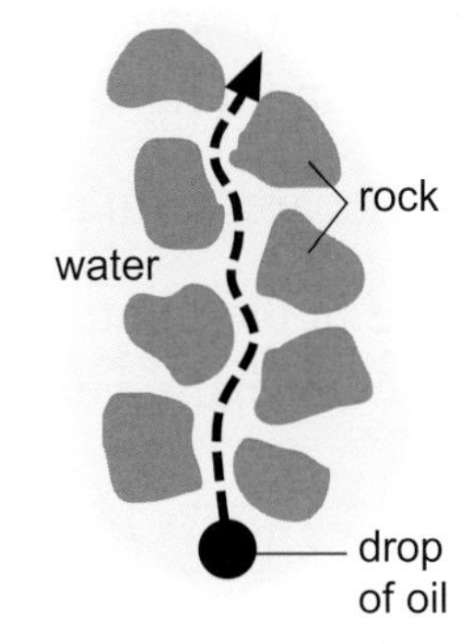

Oil drops move up though the water in porous rocks.

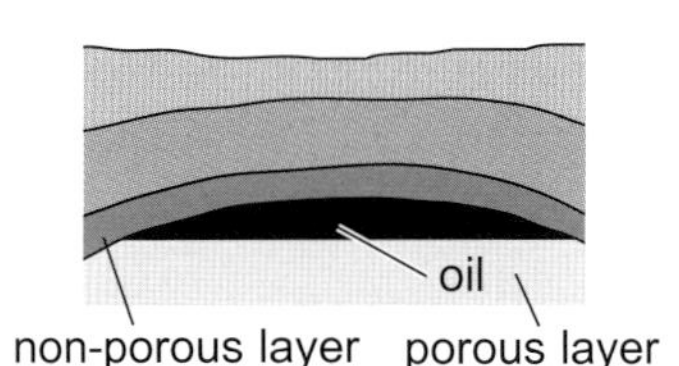

When the oil reaches a non-porous rock layer, it can't rise any further. The oil is trapped under the dome of the non-porous rock.

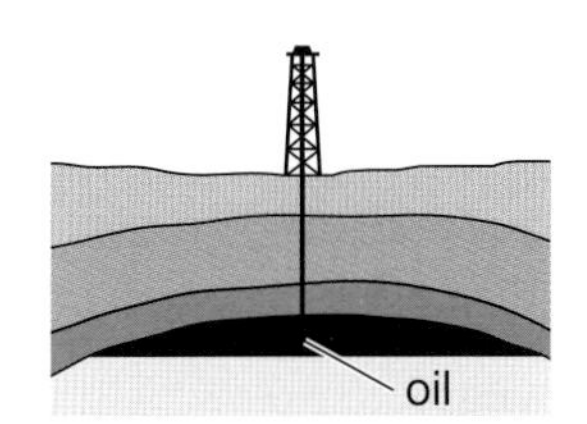

You drill through the non-porous rock to get the oil.

Rocks and minerals

The grains that make up a rock can be made from different substances. These different substances are called **minerals**. If you look closely at some rocks, you can see the different minerals.

In some rocks, the minerals are in small crystals.
In other rocks, the minerals are in much larger crystals.

Look at the picture of granite on the opposite page. It is a rock made from three different minerals. Granite is sometimes polished and used as a decorative stone because of the appearance of its minerals.

The three minerals in granite are called quartz, feldspar and mica.

Quartz	very hard, looks glassy, often a milky white colour, cannot be scratched with a knife
Feldspar	there a lots of different feldspars, often seen in pink or white crystals
Mica	small black crystals

You should already know | Outcomes | Keywords

Rocks change

Rocks do not stay the same for ever.

They get worn away very slowly by lots of different things. Rain, frost and temperature changes can all wear away rocks.

When rocks are worn away by thesc things, we call it **weathering**.

The statue in the photograph is made from carved limestone. The rain over many years has eaten away the rock.

Question 1 | 2

How does rain attack limestone?

- Rainwater is a very weak acid.
- Limestone is made from calcium carbonate.
- Calcium carbonate reacts with acid to produce carbon dioxide gas, water and a salt.
- When acid rain reacts with limestone, the carbon dioxide goes into the air and the other products are washed away by the rain.

Rain is a very weak acid so the reaction is only very slight. It takes many years until you notice the effect on limestone.

When rainwater eats away limestone, a chemical reaction happens. This is an example of **chemical weathering**.

Carbon dioxide is given off.

It is a very slow process. You can see the effect if you look at the changes in a limestone wall or in the carvings on gravestones over many years.

Chemical weathering is happening faster than it used to. The way we live now produces more gases that make acid rain.

Weather changes limestone. Acid rain makes it change even faster.

Question 3 |  4

Physical weathering

Physical weathering does not involve a chemical reaction. Rock is broken up by forces caused by changes in temperature. One way this works involves water freezing in cracks in the rock.

The water in this bottle changed to ice.

Water occupies a larger volume when it freezes. We say that it expands. (This is unusual – most substances contract when they freeze.) When water changes to ice, the **expansion** can produce very large forces.

You can show this by freezing water in a bottle. The bottle will be broken by the expanding ice. This even works for strong containers like cast iron!

The force of expansion can break off large boulders.

Rainwater gets into cracks in the surface of a rock. When the temperature drops enough, the rainwater freezes. So, the crack gets larger. This means that more water gets in when the temperature rises.

The process repeats until, eventually, a bit of the rock breaks off. Sometimes, quite large boulders break off, as in the picture.

Heat alone

The heat of the sun can make the surface of a rock get a little bit bigger. We call this expansion, too.

This rock has been cracked by hot days and cold nights in the desert.

When the temperature falls at night, the rock gets smaller again. We say that it contracts.

Repeated expanding and contracting can crack the surface of a rock – and even cause the whole rock to crack.

Other ways of changing rocks

Wind and running water can wear away rocks.

A scree slope is a lot of small rock pieces produced by weathering.

Wind blows fine particles of dust and sand against rocks. This wears away the surface of the rock.

Plants can root themselves in the cracks in rocks. The plant holds moisture. The shade from the plant also means that parts of the rock surface reach different temperatures in the sun.

Some animals like limpets make an acid that will attack rocks.

Weathering that is caused by plants or animals is called **biotic weathering**.

The holes in this rock have been worn away by sand blown by the wind.

Question 5 6 

8G.3 Moving rocks

You should already know | Outcomes | Keywords

Moving pieces of rock around

Weathering can cause rocks to break up into small bits. These are called **rock fragments**.

Three things can move rock fragments:

- gravity
- wind
- water

Gravity makes loose rock fragments fall or roll down slopes.

Wind blows tiny rock fragments from one place to another.

Rainwater washes small rock fragments down slopes.

Once rock fragments are washed into a stream or river, the small fragments get carried away and larger ones get left behind.

Question 1 | 2

In the hills, streams flow quickly. Streams carry smaller rock fragments away and leave the large fragments behind.

Further downstream, we see beaches made of pebbles.

Forming sediments

The speed of a flowing stream or river depends on things like the slope of the land. A quickly flowing river can move large rocks. Slowly flowing streams or rivers only move tiny rock fragments like fine sand.

The speed of a stream or river drops when the slope changes. This makes some of the rock fragments settle on the stream or river bed.

This is called **deposition**. The rock fragments that settle on the bottom are called **deposits**. The deposits build up to form **sediments**.

The pictures show the sediments deposited at different stages of a river's journey.

On flatter land, the river flows more slowly, so it deposits sand.

Question 3 | 4

The river deposits fine sand and mud as it gets nearer to the sea.

Changes in rock fragments

Rock fragments are made by weathering.
When they are made, the rock fragments have sharp edges and corners.

Rock fragments knock against each other as they move along. The sharp corners and edges become more rounded.

This is called **abrasion**.

The same idea is used to clean stone walls and pavements. Jets of air mixed with sand are blown at them at high speed. This is called sand blasting.

These pebbles have been worn smooth by rubbing against each other in water.

Question 5 | 6

Layers of sediment

Stir up a beaker of water that contains a mixture of different-sized rock fragments. You will see that the heavy fragments settle to the bottom first. The lighter fragments settle on top.

You get different layers forming, with the larger, heavier fragments at the bottom.

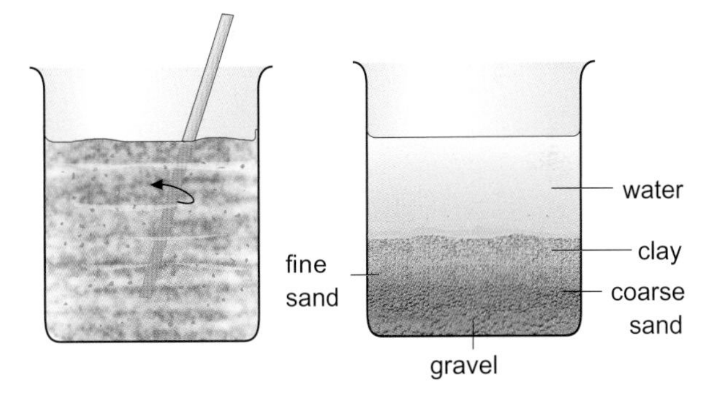

The same effect happens in nature.

Rivers deposit sediments on the sea bed. Over millions of years, there are many changes in what the rivers carry. The sediments at the bottom were deposited before the sediments that are on top.

You can see this in the picture. First, a layer of mud was deposited, then a layer of sand, then another layer of mud.

The layers build up over millions of years. The older layers get squashed and harden.

These form rocks we call **sedimentary rocks**, like mudstone and sandstone.

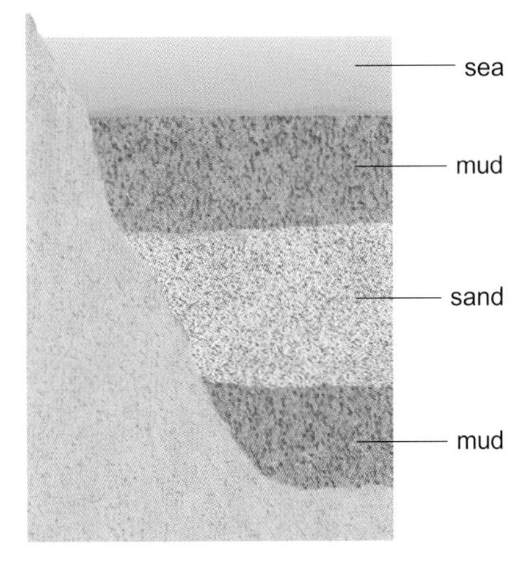

Sometimes rivers carry sand into the sea. At other times they carry mud.

Question 7 | 8

Check your progress

8G.4 More about rocks

You should already know | Outcomes | Keywords

Rocks from dissolved solids

As a result of physical weathering

- rock fragments carried by rivers form layers of deposit on the sea bed
- over millions of years, new sedimentary rocks form from these sediments.

As a result of chemical weathering

- salts dissolve in rainwater
- these salts get washed into streams and rivers.

Mineral water, showing the dissolved salts.

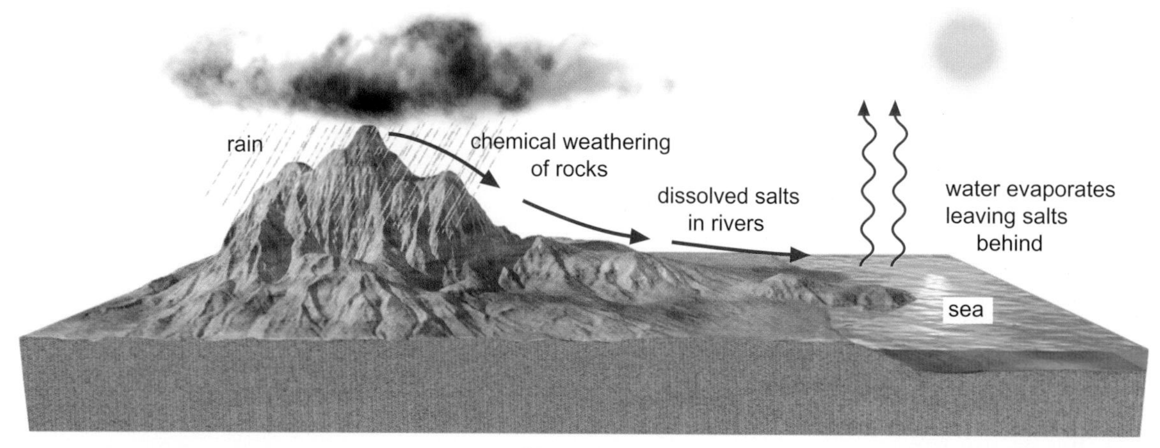

How dissolved solids become concentrated in seas and lakes.

When water evaporates from the sea or from a lake, the salts that were dissolved in it get left behind as crystals. The crystals form parts of the sediment in the sea and therefore part of sedimentary rocks.

In some parts of the Earth, the evaporation from a lake is so fast that crystals form at the edge of the lake.

Question 1 | 2

Water evaporates from this soda lake faster than it flows in. As water evaporates, crystals of salts are deposited.

Stories in rocks

When animals with shells or skeletons die, the soft parts of their bodies decay. The hard parts are left behind. They can end up in sedimentary rocks. We call these remains **fossils**.

Fossils are very common in sedimentary rocks like coal and limestone.

Fossils can help us work out how a rock formed.

Limestone contains fossils from tiny sea creatures. So limestone must be a rock that was formed under the sea.

This piece of coal has the fossilised remains of a plant in it.

Fossils in limestone.

Coal contains fossils of plants similar to ferns. This means that coal did not form under the sea. Coal formed when trees and other plants were buried by the mud in swamps millions of years ago.

We can use fossils to tell how old rocks are.

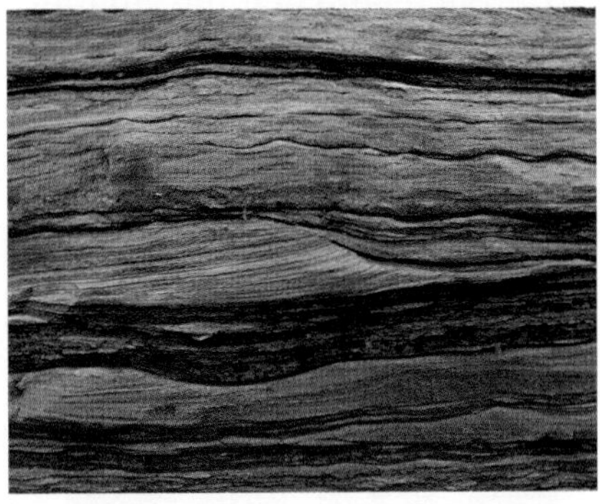

This rock formed in a river delta. The stripes in the rock are called current banding.

Question 3

The shape of rocks can show where there was running water. Some sandstone has ripple marks in it like sand on a beach.

The rocks in a cliff face take millions of years to form. The drawing shows the rocks in a cliff face. It tells us a lot.

- Shale is the oldest layer because it is at the bottom.
- The limestone contains coral fossils. At the time it was made, the sea must have been clear, warm and shallow, because those are the conditions that coral needs to grow.
- We can tell that the sediments formed for a long period of time in a clear sea because we have a thick layer of light sandstone.
- On top of the sandstone, there is a layer of salt. This means that the sea in that region must have dried out for a while.

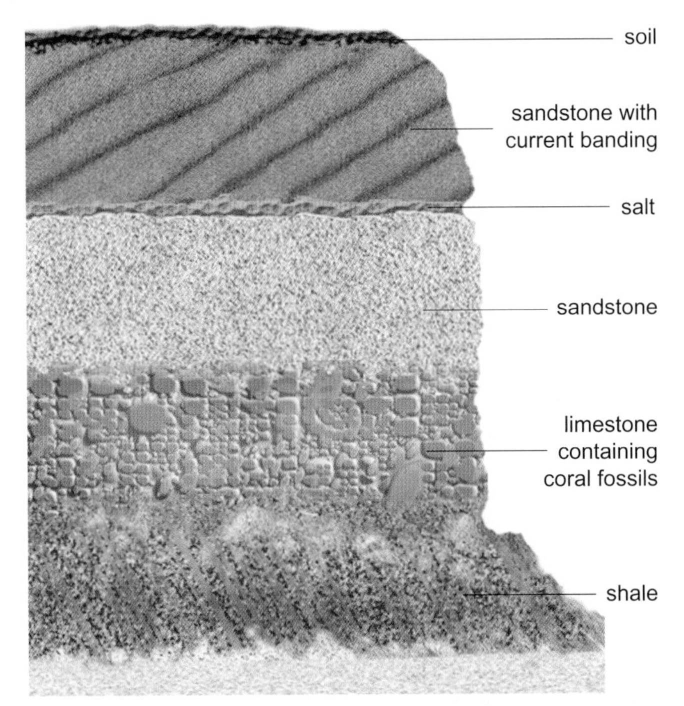

Rocks and fossils take millions of years to form. This sort of timescale is hard to imagine. Rather than use years and centuries, scientists sometimes use a scale called **geological time**, which breaks the past 600 million years or so into sections.

Review your work

Question 4 **5**

Summary ➡

8G.HSW Measurements and conclusions

You should already know

Outcomes

Keywords

Range

The **range** of measurements you make in an investigation is the difference between the smallest measurement and the largest.

It is important to make measurements over a large enough range. If your range is too small, you might miss an important result.

Two students do an experiment on a piece of limestone to see how the concentration of acid affects how long it takes to dissolve.

100cm^3 acid and 0cm^3 water
90cm^3 acid and 10cm^3 water
80cm^3 acid and 20cm^3 water
70cm^3 acid and 30cm^3 water
60cm^3 acid and 40cm^3 water
50cm^3 acid and 50cm^3 water
40cm^3 acid and 60cm^3 water
30cm^3 acid and 70cm^3 water
20cm^3 acid and 80cm^3 water
10cm^3 acid and 90cm^3 water
0cm^3 acid and 100cm^3 water

These are the acid concentrations that Lauren suggested.

80cm^3 acid and 20cm^3 water
81cm^3 acid and 19cm^3 water
82cm^3 acid and 18cm^3 water
83cm^3 acid and 17cm^3 water
84cm^3 acid and 16cm^3 water
85cm^3 acid and 15cm^3 water
86cm^3 acid and 14cm^3 water
87cm^3 acid and 13cm^3 water
88cm^3 acid and 12cm^3 water
89cm^3 acid and 11cm^3 water
90cm^3 acid and 10cm^3 water

These are the acid concentrations that Daniel suggested.

Crushed limestone dissolves faster in strong acid.

Question 1 

Consistency

Once you have carried out an experiment, you need to be sure that your **conclusion** fits in with the **evidence** from your results. We say that the conclusion and the evidence need to be **consistent**.

The experiment below investigates the effects of freezing and thawing on three different rocks. The aim is to help understand freeze–thaw weathering.

The results of the experiment are shown in the table.

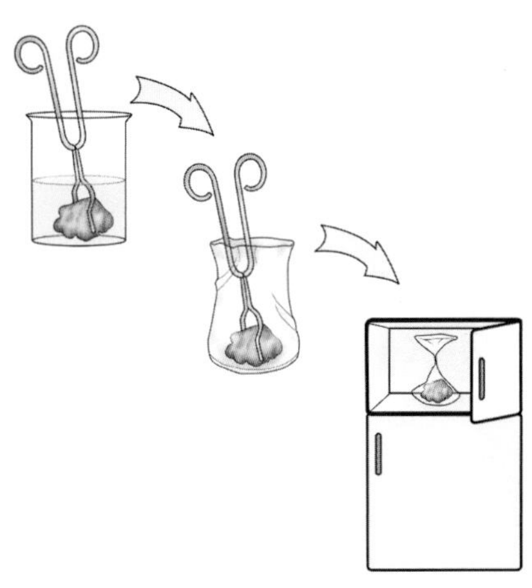

The rock is dipped in water and put in a bag in the freezer for 24 hours. The process is then repeated.

Rock	Observations		
	At start	After three freezes	After eight freezes
limestone	two small cracks	cracks have grown slightly longer	one piece has broken off, and a few more cracks have appeared
granite	no cracks	no cracks	no cracks
slate	many cracks between layers	one layer has broken off	the rock is now in five pieces

Three pupils doing the experiment think about three different conclusions.

1 Freeze–thaw weathering affects limestone the most.
2 Freeze–thaw weathering affects granite the most.
3 Freeze–thaw weathering affects slate the most.

The conclusion that is consistent with the evidence is the third one: that freeze–thaw weathering affects slate the most.

Question 3 **4**

8G Questions

8G.1

1 What is a geologist?

2 Give three examples of the different properties of rocks.

3 a What is the difference between a porous rock and a non-porous rock?

b Explain why the difference happens.

4 Describe how porous and non-porous rocks play a part in finding oil in the Earth.

5 a What is a mineral?

b Give three examples of minerals and describe where they can be seen.

8G.2

1 What is weathering?

2 What has caused the weathering of the statue in the photograph?

3 a What is chemical weathering?

b Give one example of how it can happen.

4 Cleopatra's Needle is a stone carving that was brought to London from Egypt just over 100 years ago. It had stood in the desert for thousands of years. When it arrived in London, the carvings on it were very clear. They are now very weathered.
Explain why this change has happened so quickly.

5 Describe the process of physical weathering caused by rainwater freezing in cracks.

6 a What is meant by biotic weathering?

b Give one example of biotic weathering.

8G.3

1 How are rock fragments formed?

2 Describe three ways that rock fragments are moved in nature.

3 What is meant by the term deposition?

4 a Describe how sediments form on the beds of rivers.

b Explain why some rock fragments travel further than others.

continued

5 What is meant by the term abrasion?

6 Give two examples of the use of abrasion.

7 If you have layers of different sediment, which one is the oldest? Give a reason for your answer.

8 a How do sedimentary rocks form?

b Give three examples of sedimentary rocks.

8G.4

1 Where do many rock fragments eventually deposit to form sedimentary rock?

2 a What is dissolved in rivers as a result of chemical weathering?

b How does this material transfer to sedimentary rock?

3 a What is a fossil?

b Where are fossils found?

c Give two examples of fossils.

4 How do geologists know that limestone is a sedimentary rock formed under the seabed?

5 Why do you think that scientists use a geological timescale to describe the changes in rocks rather than centuries?

8G.HSW

1 a What is meant by the word range when a set of observations are made?

b Why is it important to use a range that is large enough?

2 a What is the difference between the range of concentrations suggested by Lauren and the range suggested by Daniel?

b Why might Lauren's suggestion be a better way of showing whether the time limestone takes to dissolve is affected by the concentration of the acid?

3 What do these words mean?

a Evidence

b Conclusion

c Consistent

4 a Explain why the third conclusion is the one that is consistent with the evidence.

b Write another conclusion that is also consistent with the evidence.

8H.1 Changing rocks

You should already know | **Outcomes** | **Keywords**

You have learned how looking at rocks can tell us things about the past.

- The weather breaks down rocks.
- Rivers carry rock fragments to the sea and into lakes.
- Layers of sediment form at the bottom of the seas and lakes.
- The layers of sediment bury older layers and squash them.

These things happen all the time. They have been happening for over four billion years (4 000 000 000 years).

Rocks made from squashed sediments are called sedimentary rocks.

This sedimentary rock is made from many thin layers of sediment. The lines can be seen in the rock.

Layers of sediment form on the sea bed.

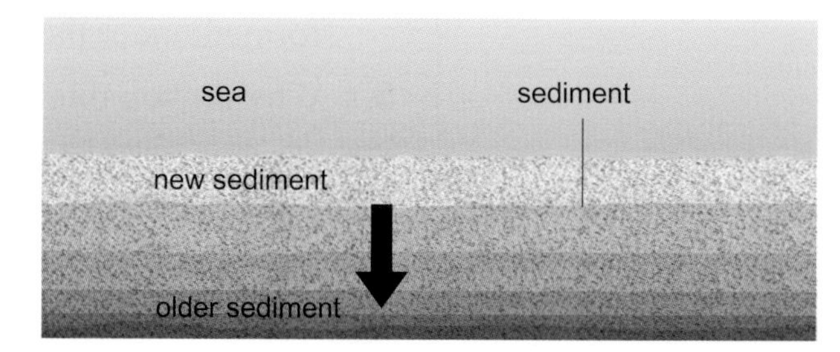

The weight of new sediments presses down on the older sediments. This pressure squeezes the water out and compresses the sediments. Chemical changes cement the fragments together, forming solid rock.

The clues that tell us a rock is sedimentary are:

- the rock has layers
- the rock contains fossils
- the rock is made of grains or particles that are stuck together
- the rock is porous.

A sedimentary rock might have only some of these clues and not all of them.

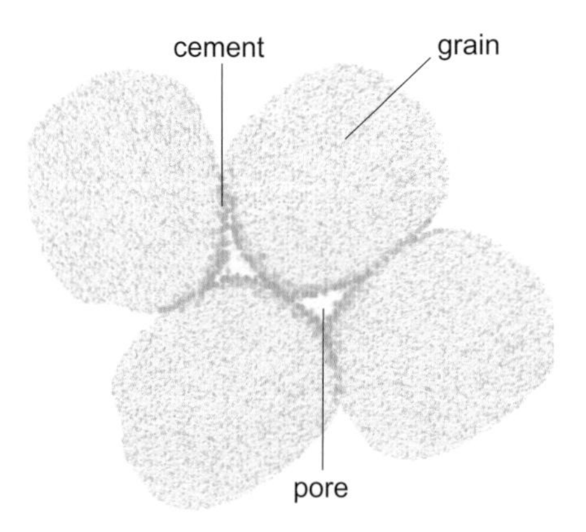

The grains of sedimentary rocks do not interlock. There are often pores or spaces between the grains. We say that sedimentary rocks are porous.

Question 1 

Changing sedimentary rocks

Sedimentary rocks get buried deeper and deeper in the Earth over millions of years. They are subjected to massive levels of **pressure** and **high temperatures**.

This can make them change their structure.

They form new types of rock.

These new types of rock are called **metamorphic** rocks.

Metamorphic rocks are usually harder than sedimentary rocks.

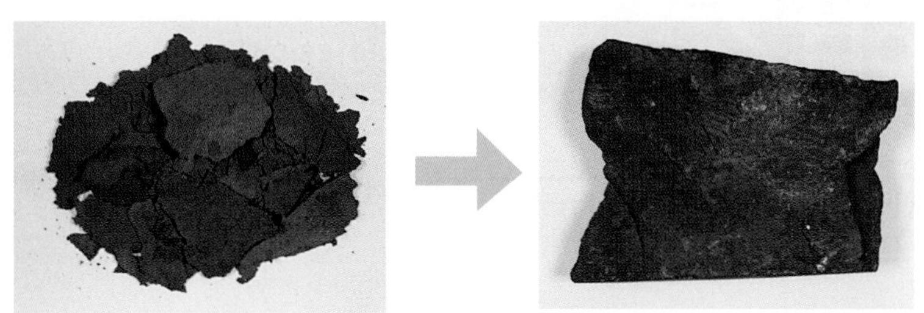

Slate is a metamorphic rock that forms from shale.

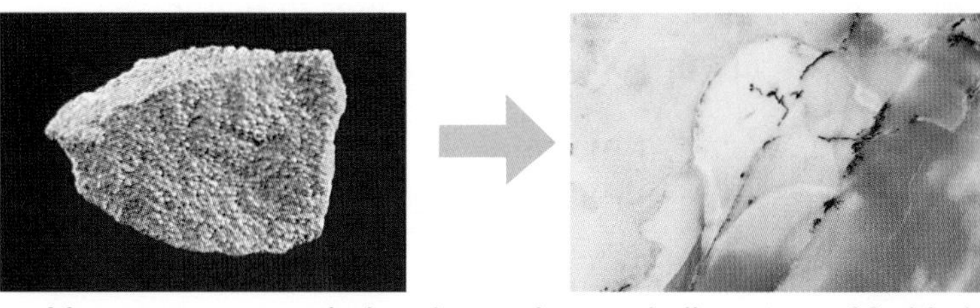

You can see rounded grains and pores in limestone. Marble is harder, with a granular, sugary texture and no pore spaces.

Question 3

4

Sandstone is made up of grains of sand. Quartzite is harder, with a sugary texture.

The heat that changes rocks comes from the molten rock called **magma** that is deep under the Earth's crust.

Sometimes this magma forces its way up into the Earth's crust.

The rocks that are near it get very hot and are squeezed. This makes them change into metamorphic rocks.

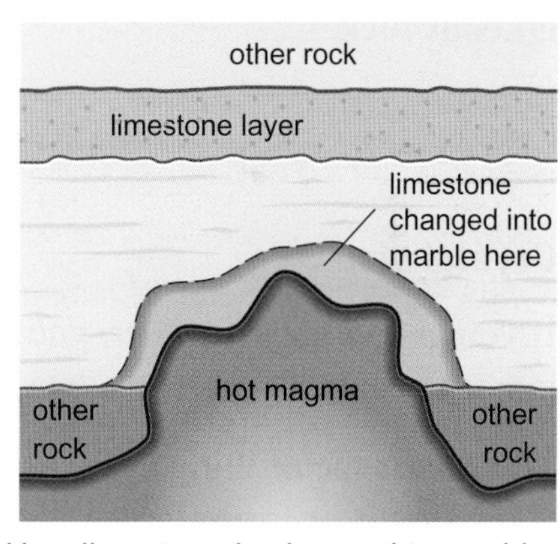

How limestone is changed to marble.

The Earth's crust is always moving very slowly. Over millions of years, the layers of rock in the Earth's surface get twisted and squashed. This produces a lot of pressure. You can sometimes see the effect on layers of rock on exposed cliff faces.
As you go deeper under the Earth's surface, the pressure and temperature increase.

This South African gold mine is so hot that the miners can only work for a few hours at a time.

Molten rock heats up the rocks near to it.

Movements of the Earth's crust cause heating and squashing of rocks.

Question 5

6

8H.2 All about magma

You should already know | Outcomes | Keywords

When solids get hot, they melt and change into a liquid.
This happens at 0 °C when ice changes to water.
It happens at about 1000 °C for rocks like granite.

Liquid rock is called **magma**.
A large part of the middle of the Earth is made of this molten rock.

The crust of the Earth is made of solid rock. If the Earth was the size of an orange, the crust of solid rock would be like the orange peel.

When magma cools, it forms solid rock. Magma is a mixture of different minerals, so it often forms different crystals when it solidifies. You can see the crystals in the rocks that form. Look at the photographs below.

When a rock is made by magma cooling into a solid, it is called an **igneous** rock.

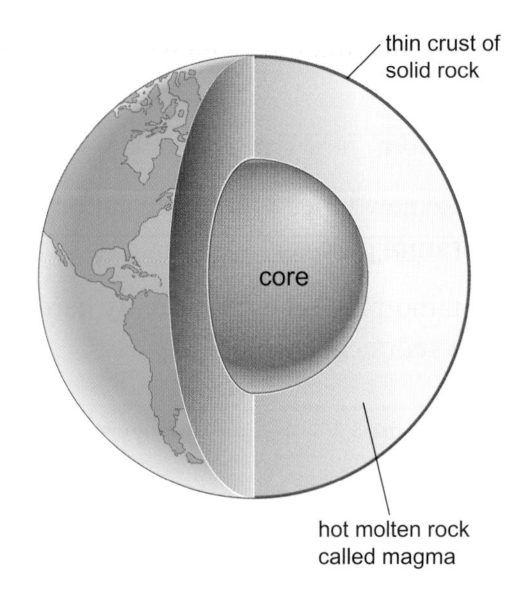

Cross-section through the Earth.

Question 1

Crystal size

If you evaporate salt water, you get salt **crystals**. You get much larger crystals if you let the water evaporate slowly.

The same type of thing happens when magma cools. If magma cools slowly, it forms large crystals. If magma cools quickly, it forms small crystals.

these crystals formed quickly

these crystals formed slowly

The size of the crystals depends on how quickly you evaporate the water from the solution.

Granite.

Basalt.

Pumice is an igneous rock that has cooled so fast that you can see gas bubbles but no crystals.

Question 2

Some magmas cool more slowly than others

Magma cools to make igneous rock in three ways.

- The magma cools deep underground.
  Because it is surrounded by hot rock, it cools very slowly.
  It will form large crystals.
- The magma forces its way nearer to the surface through cracks in rocks.
  This magma cools more quickly. It forms medium sized or small crystals, depending upon how fast it cools.
- It comes out of a hole in the Earth's surface as a liquid.
  This magma cools rapidly. It will form very small crystals or even not form any crystals at all.

Lava from a volcano in Hawaii.

Volcanoes

We call magma that reaches the Earth's surface **lava**.
When lava emerges on the Earth's surface, we say that it **erupts**.

Mount St Helens is an ash cone. Ash was blown out in a cloud.

Some lavas are very runny and spread over a large area.
Other lavas are less runny and build up a cone of rock. This is a type of **volcano**.

Sometimes the magma solidifies inside a volcano. It acts like a giant plug. Gases are trapped under the plug. The pressure can build up until the plug cannot hold back the material underneath.

When this happens there is a violent eruption. Gases, solids and molten lava burst out of the top of the volcano like froth from a shaken fizzy drink.

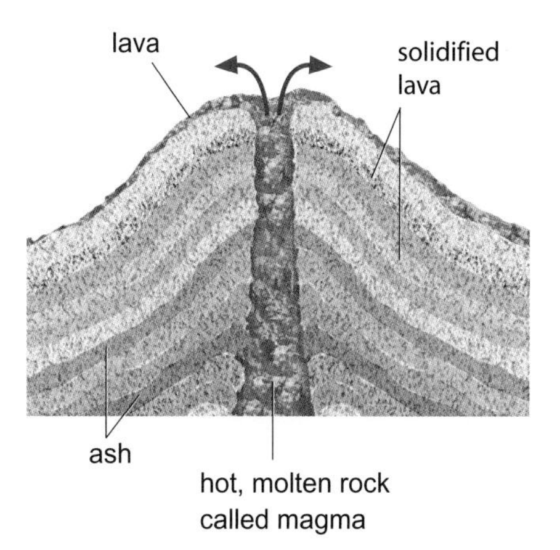

Some volcanic cones form from ash and lava.

This type of eruption produces a lot of volcanic ash, which can form a large ash cone.

This type of eruption happened at Vesuvius in Italy in AD 79.
It buried the nearby town of Pompeii in volcanic ash.

Question 3 4

The volcano in the diagram has formed by layers of lava emerging and solidifying, followed by eruptions of ash, all forming a cone of new rock.

Volcanoes can happen under the sea. The island of Surtsey is off the south coast of Iceland. It formed in 1963 when a volcano erupted. It is now a nature reserve.

Surtsey erupted under the sea and formed a new island.

Question 5

Check your progress

8H.3 Recycling the rocks

You should already know | Outcomes | Keywords

Three main types of rock

There are three main types of rock.

- Sedimentary
  Formed when sediments are compressed and cemented together by pressure under the sea bed.
- Metamorphic
  Formed when rocks are subjected to high levels of heat and pressure deep inside the Earth's crust.
- Igneous
  Formed when liquid magma cools and solidifies into rock.

The weather can make rock fragments from all three types of rock.

Rock fragments are moved by gravity, wind and water until some of them end up as sediments in the sea. They make new sedimentary rock.

As new rocks are made, the older rocks sink deeper in the Earth's crust. Some are changed into metamorphic rock by heat and pressure.

The deepest and oldest rocks melt into the magma underneath the Earth's crust.

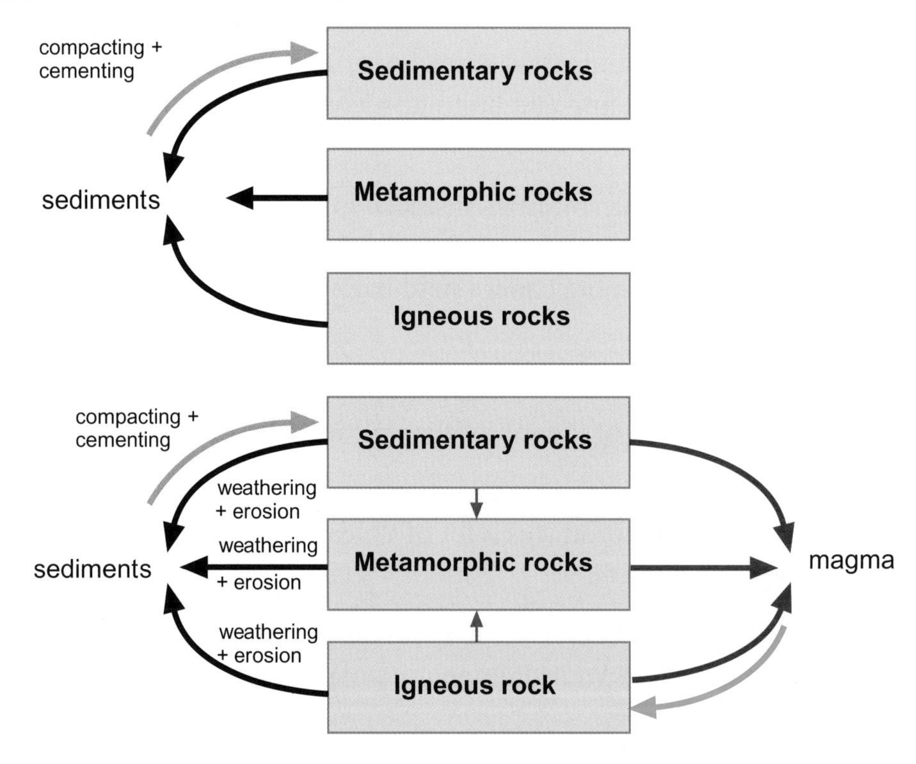

The material from those rocks can be recycled into new igneous rock when any of the magma cools and solidifies. The rocks may sooner or later arrive on the Earth's surface.

This constant recycling of the material in rocks is called the **rock cycle**.

Question 1 | 2

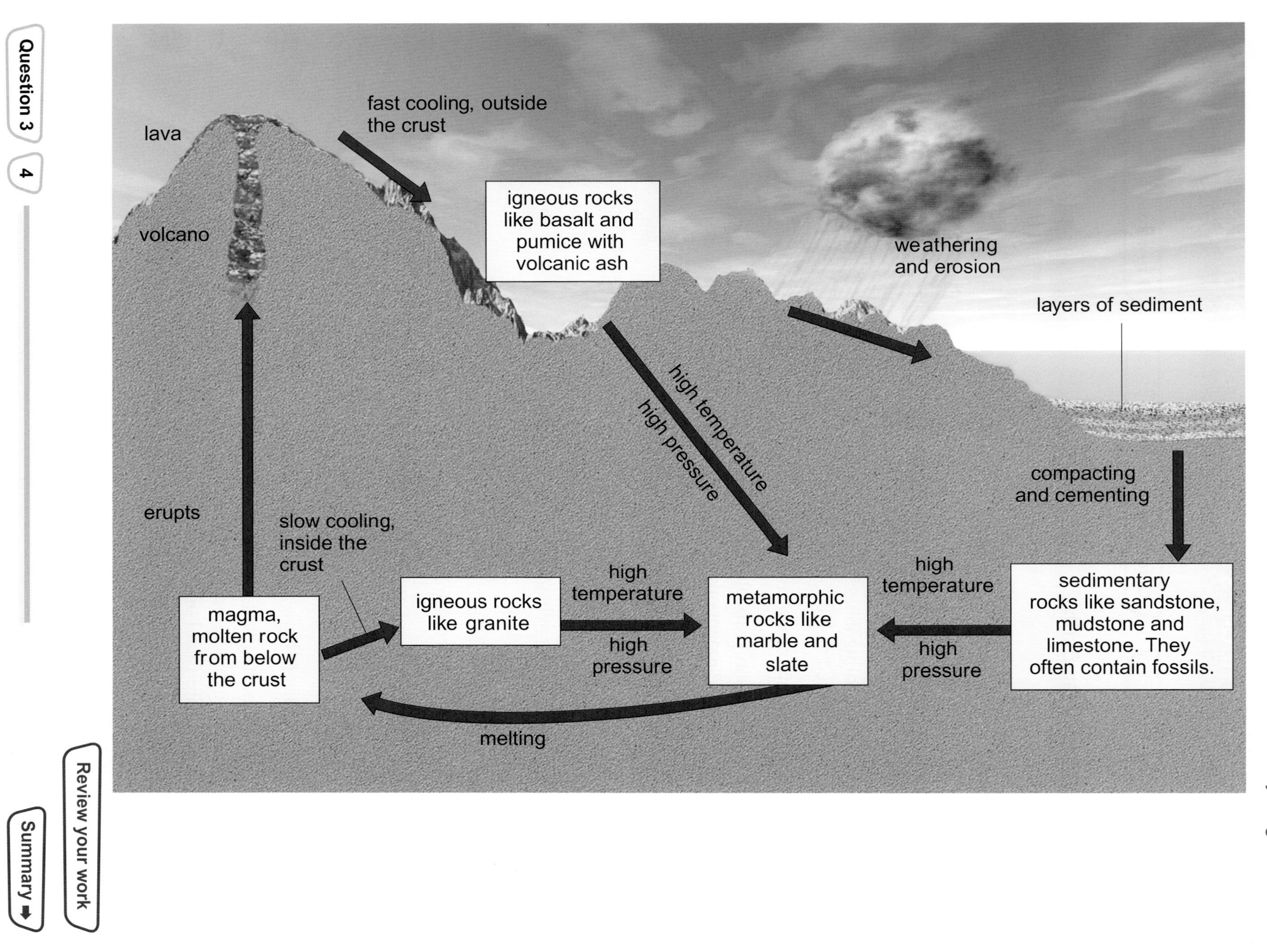

Review your work

Question 3 | 4

Summary ➡

8H.HSW Theories about the structure of the Earth

You should already know | Outcomes | Keywords

Changing ideas about the Earth

People used to believe that

- the Earth was only a few thousand years old
- the Earth never changed.

Geologists worked out that the Earth was much older. The next theory was that

- the Earth formed from cooling molten rock
- the Earth's surface wrinkled as it cooled.

This theory explained mountains and valleys.

The Earth is over 4000 million years old. If it had been cooling for long enough to create the mountains by shrinking, then it would have cooled so the middle was completely solid. We know this is not true. The shrinking Earth theory must be wrong.

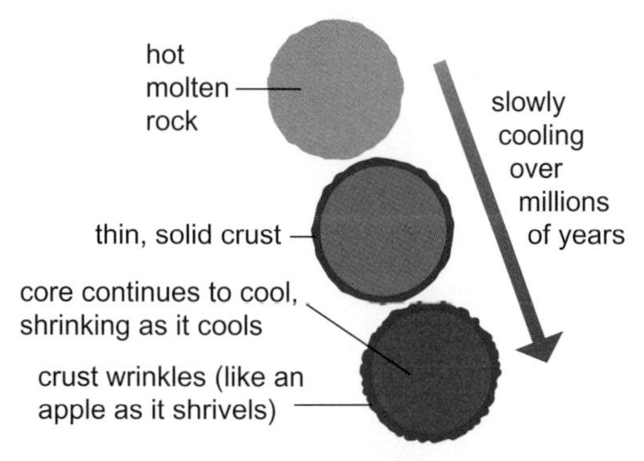

The shrinking Earth theory.

crust is solid rock

mantle of solid rock

molten outer core

solid inner core

both made of the metals iron and nickel

Scientists think this is what the Earth is like inside. This model explains all the observations.

The idea of a moving crust

Vibrations from earthquakes and explosions travel through the Earth. Scientists use these to work out what the Earth must be like inside.

In 1912, a German scientist called Alfred Wegener suggested that

- millions of years ago, all the land was in one large continent
- over millions of years, this land split and moved apart.

This is sometimes called **continental drift**.

Wegener had some evidence to back up his ideas.

- The shapes of the continents fitted together.
- When they were fitted together, the places where particular rocks are found on different continents line up.
- Similar fossils are found on different continents where they fit together.

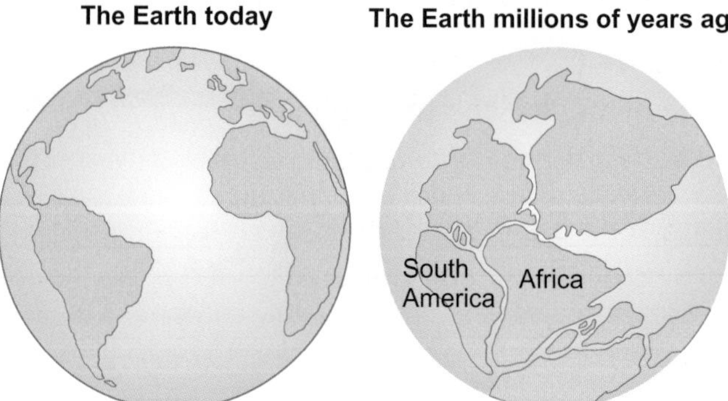

We think the continents we know today all broke away from a larger land mass millions of years ago.

Question 1 | 2

Another theory

Wegener's idea was not accepted by everyone because he could not explain why the continents had drifted apart.

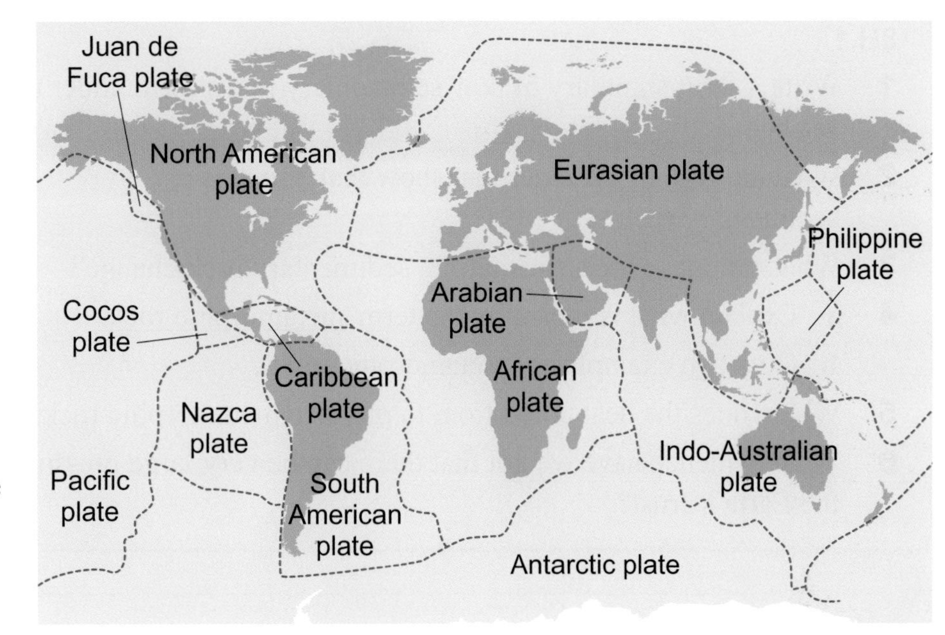

Tectonic plates.

In the 1960s, a new theory was developed. This explained observations of rocks on the ocean floor. The rocks have patterns of magnetism in them. To explain this, geologists have made up a theory of **tectonic plates**.

According to this theory:

- the Earth's surface is made of several large plates
- the plates are slowly moving about on the molten magma underneath them
- these plates can move up to a few centimetres in a year.

This theory explains Wegener's idea of continental drift. It also explains some other things.

The places where the plates meet are called **plate boundaries**. This is where the plates are moving apart or pushing together.

This can cause sudden movements, which produce earthquakes and volcanoes.

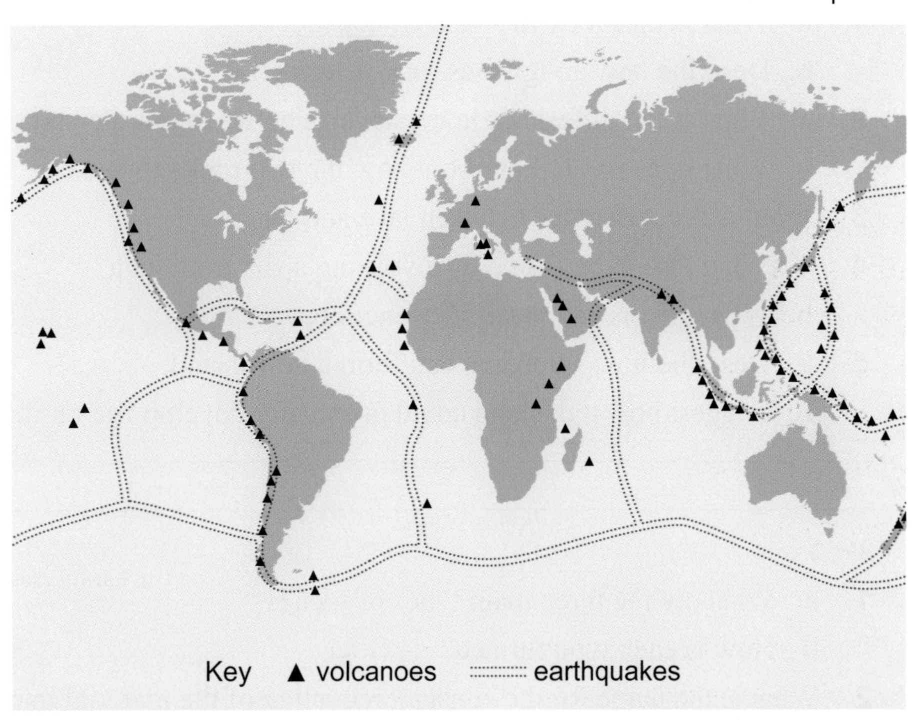

Where most earthquakes and volcanoes happen around the world.

If you look at a map of the Earth and plot where earthquakes and volcanoes happen, the sites match the plate boundaries.

Question 3 

8H Questions

8H.1

1 Write a short summary of how sedimentary rocks form under the sea bed.

2 What are four of the clues that show that a rock is a sedimentary rock?

3 What can make the structure of a sedimentary rock change?

4 **a** Explain what is meant by the term metamorphic rock.
 b Give two examples of metamorphic rocks.

5 Where does the heat come from to produce metamorphic rocks?

6 What evidence have we got that there can be very large pressures in the Earth's crust?

8H.2

1 **a** What is meant by the term igneous rock?
 b Describe how an igneous rock is formed.

2 **a** What is the difference in crystal size between granite and basalt?
 b What does this tell us about how the two rocks formed?

3 What is the difference between lava and magma?

4 **a** What else is produced by a volcano apart from lava?
 b What effects can these substances have?

5 **a** Describe how a volcano can form a new island.
 b Suggest how the island could eventually develop life on it.

8H.3

1 **a** What are the three main types of rock?
 b How is each type formed?

2 What is the name for the constant recycling of the material in rocks?

3 Which rocks melt into the magma in the rock cycle?

4 Describe the journey of a mineral from a piece of granite on a cliff over millions and millions of years as it gradually travels through each type of rock.

8H.HSW

1. How did the shrinking Earth theory explain the appearance of mountains and valleys on the Earth's surface?
2. What evidence did Wegener have to support his idea of continental drift?
3. What is the idea that is now used to explain continental drift and the location of earthquakes and volcanoes?
4. Find out what the San Andreas fault is and why people who live in San Francisco are concerned about any tectonic plate movements.

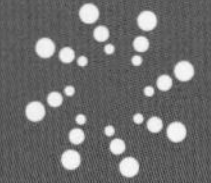

Scientific enquiry

8I.1 Measuring how hot things are

You should already know

Outcomes

Keywords

Temperature

You feel how hot or cold things are with your skin.

There is a problem with this. If you move from a hot place to a cold place, you will think that it is colder than it actually is.

We call the idea 'how hot or cold something is' **temperature**.

To get an accurate temperature measurement, you must use a **thermometer**.

Question 1 **2**

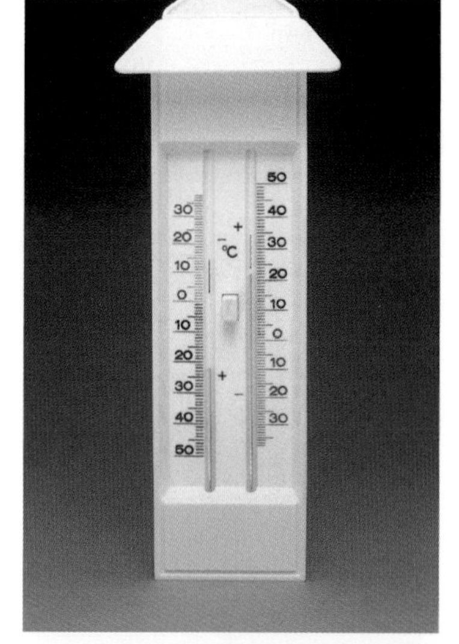

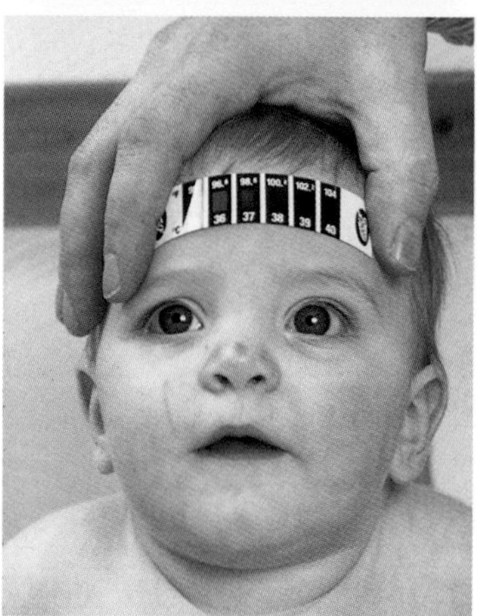

Two different types of thermometer.

Types of thermometer

There are many different types of thermometer.

Some use a liquid that expands and contracts when the temperature changes. These are called liquid-in-glass thermometers.

Many thermometers are digital. These are easier to read and less fragile than liquid-in-glass thermometers. They cost more and need a battery or power supply.

The first type of thermometer was invented by Galileo in 1592. It used the expansion and contraction of air to show the temperature. It was not very accurate.

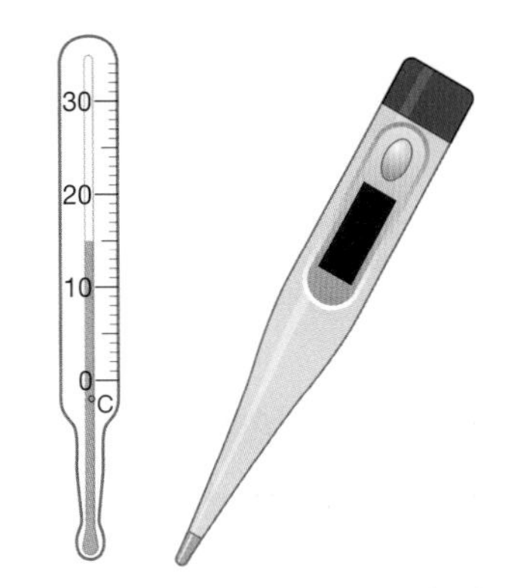

A liquid-in-glass and a digital thermometer.

Galileo's thermometer.

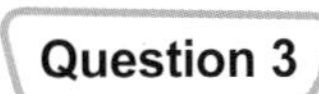

Question 3

Temperature scales

Many different scales have been used to measure temperature. The scale we use today is called the **Celsius** scale.

Some countries still use a scale called the Fahrenheit scale. If you listen to weather forecasts or read a cookery book, the temperature is sometimes given in both Celsius and Fahrenheit.

A temperature scale gives the temperature in divisions called **degrees**.

The degrees on the Fahrenheit scale are smaller than the degrees on the Celsius scale.

A much more important scale is sometimes used in science. It is called the Kelvin scale. This has the same size of degree as the Celsius scale.

Situation	Celsius temperature	Fahrenheit temperature	Kelvin temperature
lowest possible temperature	−273 °C	−459 °F	0 K
ice melts	0 °C	32 °F	273 K
human body temperature	37 °C	98.4 °F	310 K
water boils	100 °C	212 °F	373 K

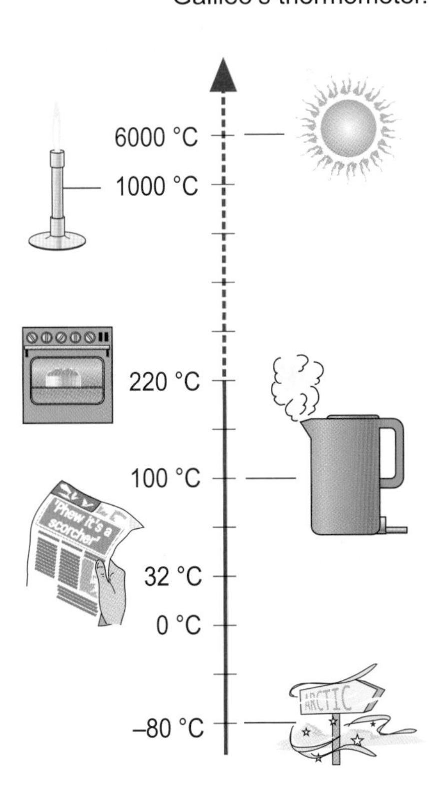

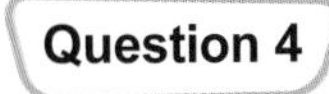

Question 4

5

You should already know

Outcomes

Keywords

Moving energy

The energy needed to make the temperature of something rise is often called **heat energy**.
Another name for this is **thermal energy**.

Heat energy moves from high-temperature places to low temperature places.

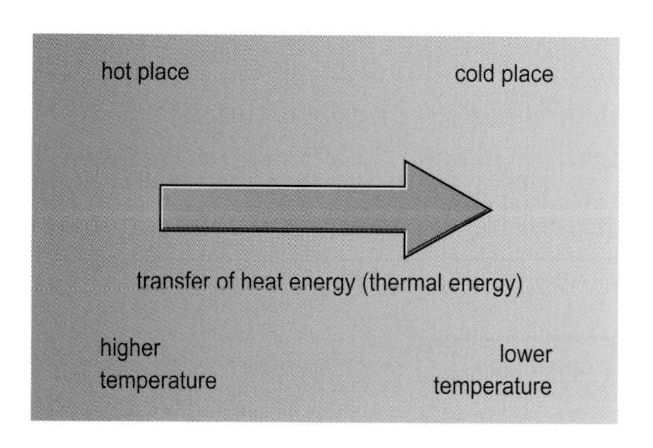

This is why

- we get heat from the Sun
- a fire heats a room
- food cooks in an oven
- we feel cold when the temperature of the air drops in winter.

Heat energy (thermal energy) is measured in **joules**. A kettle supplies about 2000 joules of heat energy (thermal energy) to the water inside it every second it is on.

The heat energy (thermal energy) makes the particles in the water move faster. The temperature is a measure of how fast the particles are moving.

A thermometer is like a speedometer.

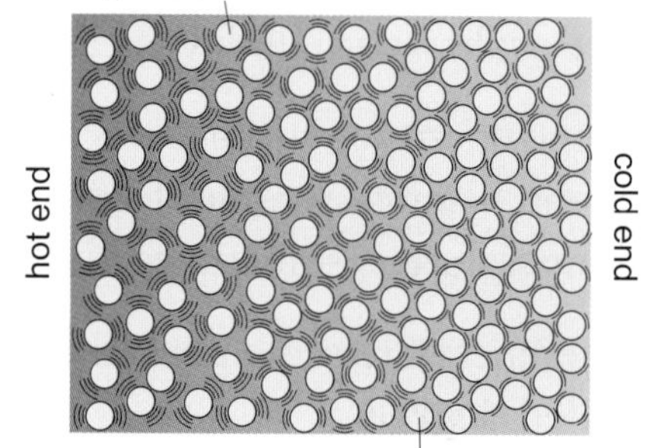

Particles in a solid with one end cold and one end hot.

Question 1 2 3

Conduction

If you connect a hot place and a cold place using some solid material, heat will travel through the solid. This is called **conduction**.

Heat passes very easily through metals. We say that they are good conductors of heat or good **thermal conductors**. Very good conductors include copper, aluminium and silver.

The diagram shows how a good conductor is used to cook baked potatoes more quickly than just putting them in the oven.

It takes a long time for baked potatoes to cook through to the middle. Using aluminium spikes, they cook through in half the time.

Question 4 5

Poor conductors

Wood and plastic do not let heat travel through them easily. We call them good **thermal insulators** or poor conductors of heat.

Hot pans or dishes should be put on mats so that the table doesn't get scorched or burned by the pan. The mat must be a good insulator.

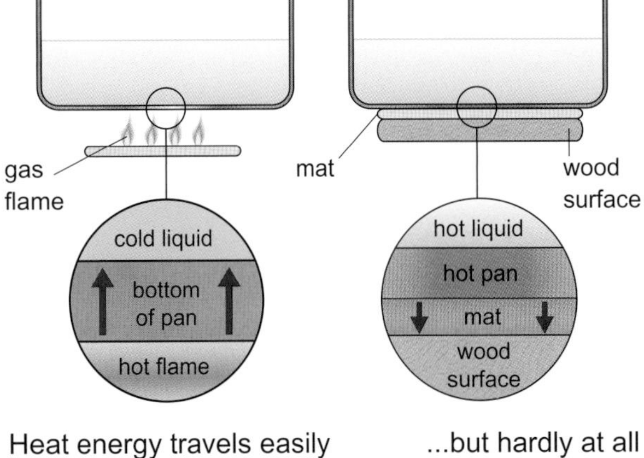

Heat energy travels easily through a conductor... ...but hardly at all through an insulator.

One of the best insulators is air. The particles in air are spread out. They only bump into each other occasionally. Most of the time, there is no connection between them. This means that the particles cannot pass the heat energy on easily. Air is a poor conductor of heat. To use air as an insulator, we need to trap it.

Insulation	How is the air trapped?	Typical uses
foam insulation	as bubbles inside plastic	packaging for fresh food that needs to stay cool
fibre insulation	as bubbles between fibres	knitted clothes, loft insulation
double glazing	as a layer between two panes of glass	windows in houses and offices

Question 6 **7**

Conduction makes some things feel cold

Sometimes, two things that are at the same temperature feel like they are at different temperatures. A good example of this is a bike with metal handlebars and a plastic foam saddle. The metal handlebars feel colder than the seat.

Metal is a good conductor. It takes the heat away from your warm hand quickly. It feels cold.

The seat is a poor conductor. It does not take the heat away quickly. It does not feel as cold.

Anna notices that the handlebars of her bicycle feel colder than the saddle.

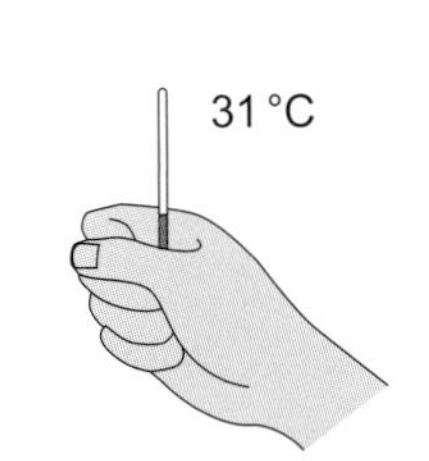

Anna's science teacher lends her a thermometer. Anna's hand is at 31 °C.

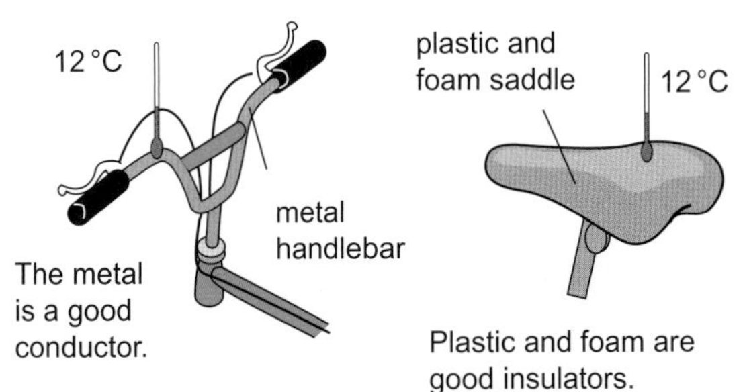

Anna measures the temperature of the handlebars and the saddle. They are both 12 °C.

Question 8

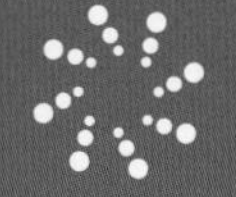

8I.3 Other ways of moving heat

You should already know | Outcomes | Keywords

You can get heat energy to move through liquids and gases, but it happens in a different way to the conduction of heat through solids.

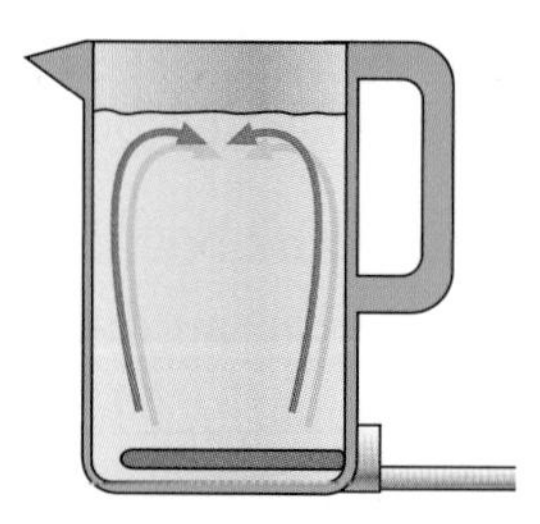

Water next to the heating element gets hotter. This hot water rises.

Boiling a kettle

Look at the pictures of a kettle.

- The heating element of a kettle heats the water around it.
- The particles in the hot water move faster and move further apart.
- The hot water is less dense than the cold water around it.
- The hot water rises up through the cold water.
- Cold water replaces the hot water around the heating element and starts to heat up.
- The water cycles around the kettle until it boils.

This is called **convection**.

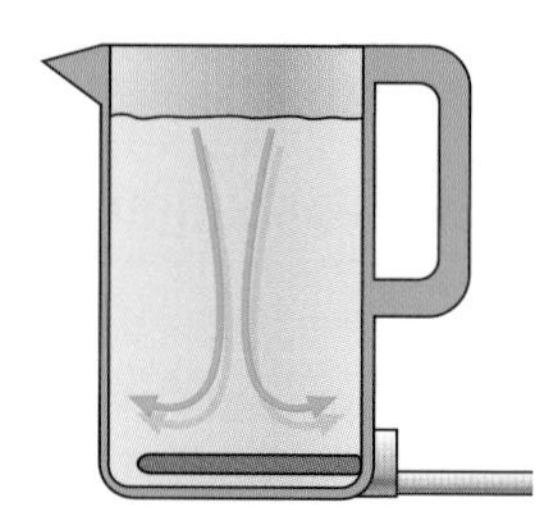

Colder water then falls down to take its place.

Gases and liquids are both fluids. They will flow. The same idea about convection applies to the air in a room. Hot air is lighter than cold air, so it rises up from around the heater. The cold air moves in to replace it. The cycle of air is known as a **convection current**.

Heat transfers in this way in liquids and gases – because they are fluids, they can flow.

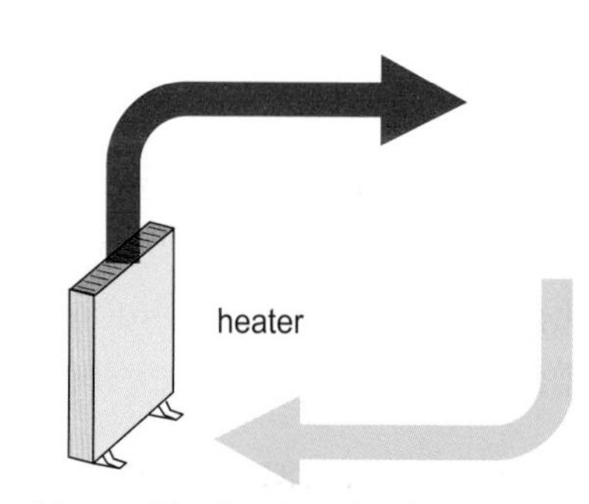

Air next to the heater becomes hotter. This hot air rises. Colder air then falls to take its place.

Question 1

Using convection currents

Convection currents occur in nature. Dark areas like farm buildings, ploughed fields and tarmac are heated up by the Sun more than light areas or green fields.

Convection currents rise above them. We call these convection currents thermals. Glider pilots use thermals to gain height.

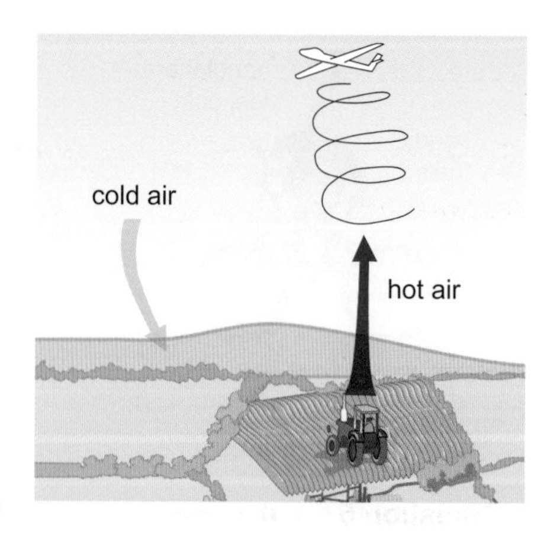

Question 2 | 3

Radiation

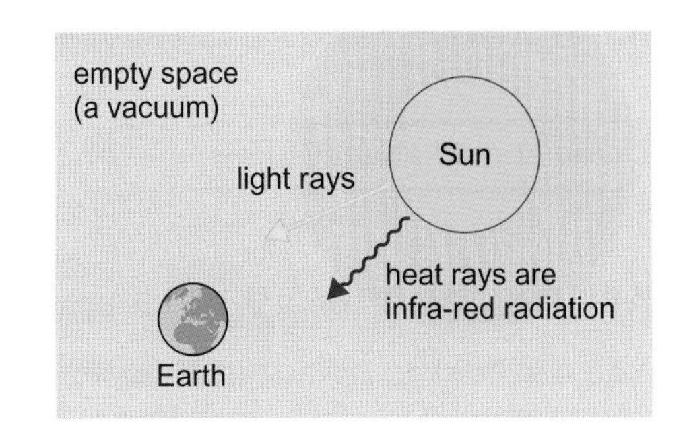

Heat energy reaches the Earth's surface from the Sun. The energy travels across 150 million kilometres of empty space.

This does not happen by conduction or convection. There is nothing between the Sun and the Earth to allow conduction or convection to happen. This method of heat transfer is called **radiation** or **thermal radiation**. Sometimes we use the full name: **infra-red radiation**.

Dark, dull surfaces are good absorbers of infra-red radiation.
This means that they soak up infra-red radiation well.
Light, shiny surfaces do not absorb infra-red radiation very much.
They are good reflectors of radiation.

Dark clothes make you feel hot on a sunny day.

Astronauts wear shiny suits for space walks.

The tar on roads can melt in the summer sun.

Houses in hot countries are often white.

Question 4

Concentrating the radiation from the Sun

You can focus the radiation from the Sun using a magnifying glass. You can concentrate the heat energy enough to burn paper.

You can also focus the infra-red radiation using a curved mirror. This idea is used in solar cookers, which can be used in sunny places like Africa, where there is no wood to use as fuel.

Heating impure water above 65 °C for about ten minutes will kill any bacteria and make it safe to drink. A solar cooker can let people do this cheaply.

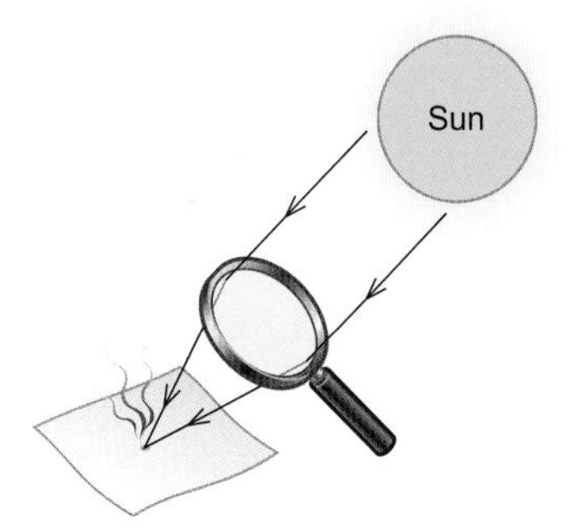

A lens focuses sunlight to a bright spot. The concentrated energy can set fire to paper.

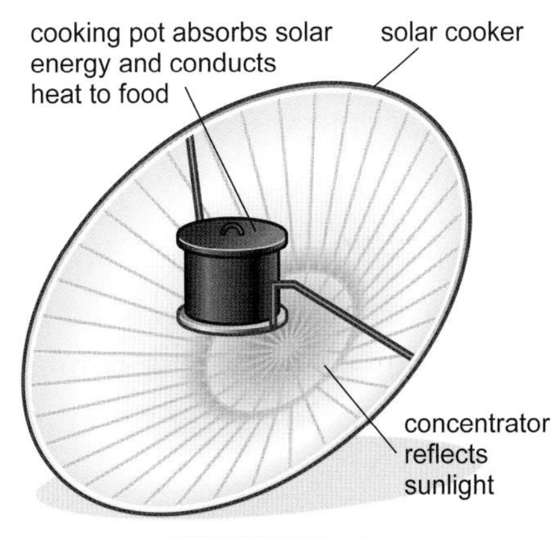

Everything gives off some level of radiation

Things at a high temperature give off a lot more radiation than things at a low temperature.

You can use the thermal radiation that is given off to take 'heat' photographs of objects. Rescue workers use infra-red cameras after earthquakes to see if people are still alive under rubble.

Question 5  **6**

Check your progress

You should already know | Outcomes | Keywords

Keeping the house warm

You can cut the cost of heating a house by half if you control the heat loss to the outside air.

This diagram shows how much energy you can lose every second from a house on a winter's day.

The total number of joules shown on the diagram would be enough to boil a 1.5 litre kettle of cold water in less than one minute!

You can insulate your loft with layers of glass fibre. Glass fibres trap air. This is a good insulator and reduces heat loss by conduction.

It costs about £300 to insulate a loft. In one year, you could save about £150 on your heating bills.

Draught excluders are very cheap. They can save about £50 in a year.

Double glazing traps a layer of air between two sheets of glass. The air acts as an insulator.

Double glazing is expensive. It takes about ten years for someone to save enough on their heating bills to cover the cost of double glazing.

Double glazing also reduces the level of sound coming in from outside. Many people have double glazing fitted if they live near an airport or motorway. It makes it quieter inside the house.

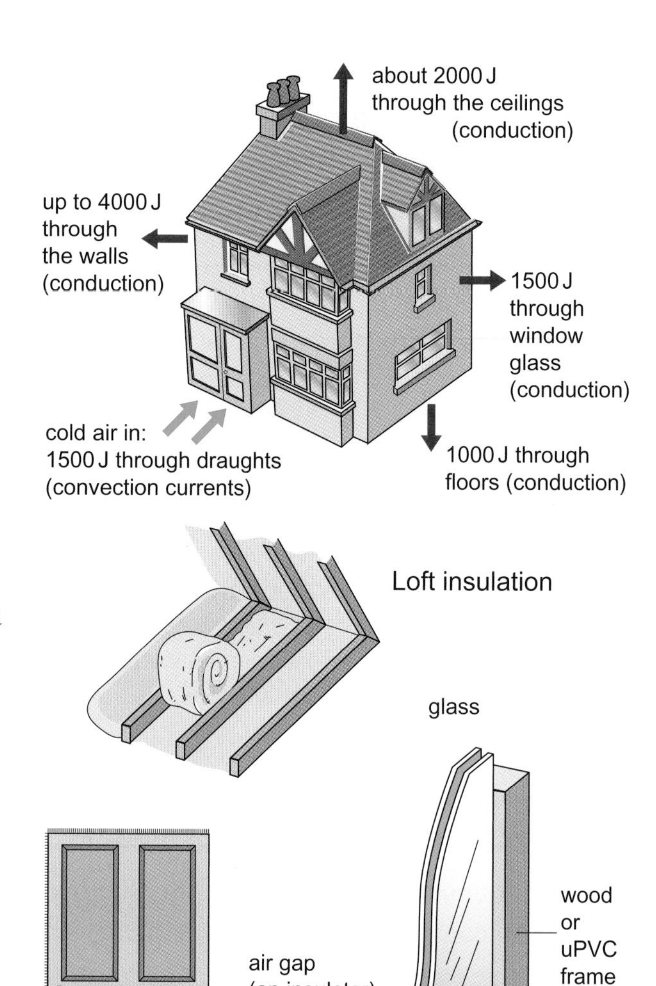

Draught excluders

Double glazing

The radiated heat from the back of a radiator can be reflected back into the room by a shiny plastic sheet on the wall.

The walls of most modern houses are made of two layers of brick with plastic foam between them. The foam contains air bubbles. This makes it a very poor conductor of heat. This is known as cavity wall insulation.

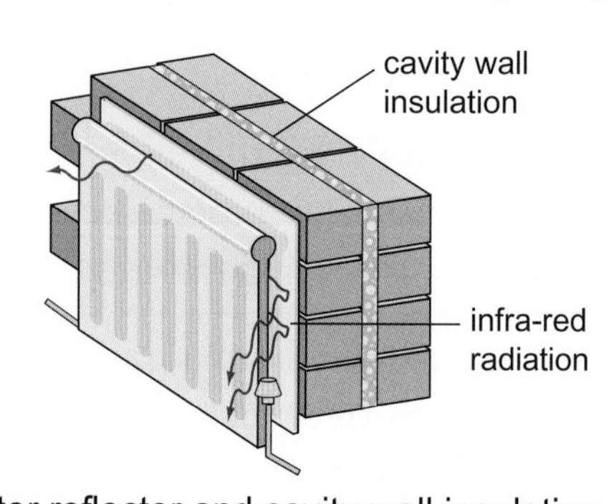

Radiator reflector and cavity wall insulation.

Question 1 

Shiny surfaces are poor radiators of heat

If you fill a black can and a white can with boiling water and leave them to cool, the black can cools more quickly.

The diagrams show the temperatures after several minutes.

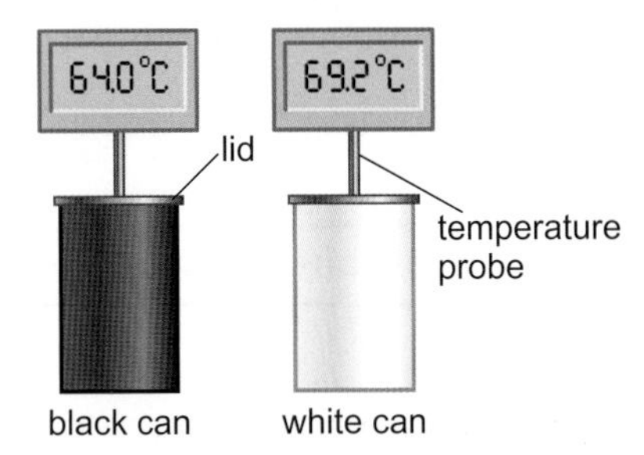

One can is painted black, one white. Otherwise, the two cans are identical.

The heat radiation given off by something depends on what its surface is like. Light-coloured, shiny surfaces give off a lot less radiation than dark, dull surfaces. A light, shiny surface is a poor **emitter** of radiation.

At the end of marathon races, runners are at risk of their body temperature falling too much. This is called hypothermia. It happens because they cool quickly in thin clothes and because of the sweat on their skin.

Runners are wrapped in shiny foil blankets after a long race to prevent hypothermia. The shiny surface on the inside of the blanket reflects heat back to the runner. The shiny surface on the outside reduces the heat lost by the runner to the outside.

The silver blanket reflects heat back to the runners' bodies and reduces heat loss by radiation.

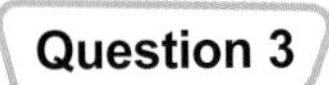

The vacuum flask

A vacuum flask (or Thermos® flask) keeps hot drinks hot and cold drinks cold.

It has a double wall with a vacuum in the gap. A vacuum is a space where all or most of the air particles have been sucked out. There are virtually no particles in a vacuum so conduction and convection cannot happen.

Part	What it does
glass wall	Glass is a poor conductor. It reduces heat loss by conduction.
shiny surfaces	The inside shiny surface does not radiate much heat. The shiny surface on the inside of the outer wall reflects any radiation that does cross the vacuum.
vacuum	Prevents conduction and convection.
stopper	Prevents convection out of the top of the flask.
foam plastic	Reduces conduction through the stopper.

Question 4

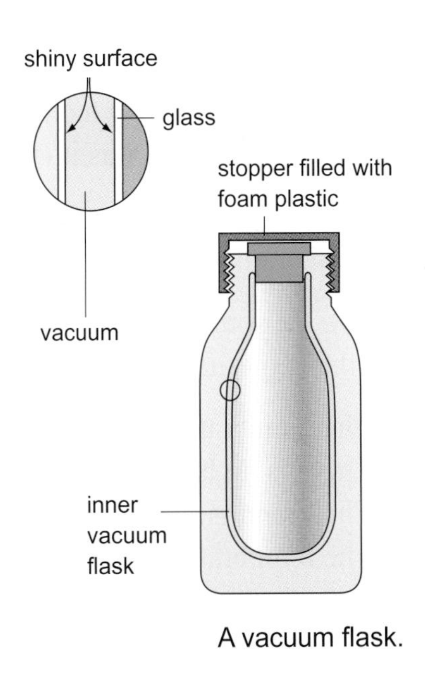

A vacuum flask.

Review your work

Summary ➡

8I.HSW Collecting data

You should already know

Outcomes

Keywords

Gathering evidence

We use our senses to observe things.

In science, we use measuring instruments to extend our senses.

For example, you can use

- a ruler to measure the edge of a book
- a scale under a microscope to estimate the thickness of a hair.

Sometimes, scientists want to measure many results very rapidly or to take a lot of results over many hours.

One way to do this is with a sensor linked to a computer. This is also known as a **datalogger**.

We can measure

- manually
- using sensors
- using sensors linked to computers

Some pupils are using a datalogger to study the temperature of water as it is heated from ice to boiling point.

The diagram shows the experimental set-up.

These are some advantages of using a datalogger for this experiment.

- The temperature sensor is accurate. There is no human error in reading the result.
- The graph is plotted as the experiment goes along.
- The temperature is measured about 60 times every second, so the graph looks like a continuous line.
- The computer stores the results. They can be used again later without having to type them in.

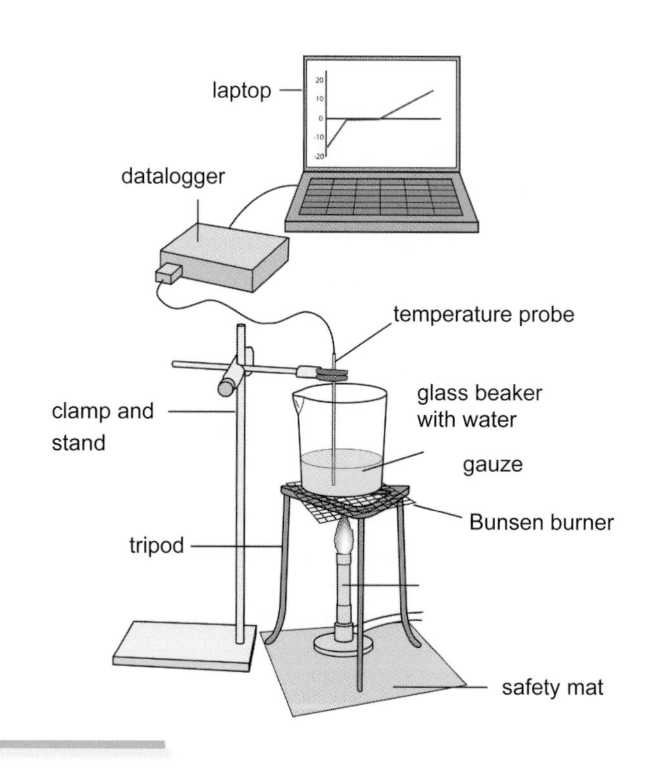

Question 1

Another advantage of a datalogger

The photograph shows a datalogger with three sensors.
These are being used to make measurements in a pond. The sensors record different things:

- the amount of oxygen dissolved in the pond water
- the temperature of the pond water
- the amount of light falling on the pond water.

The purpose of the experiment was to see whether the level of the oxygen in the pond water was affected by the level of light falling on the pool and the temperature of the water.

To carry out the experiment, it was necessary to take readings over a period of 24 hours. The graph shows the readings from the temperature sensor over a period of 12 hours.

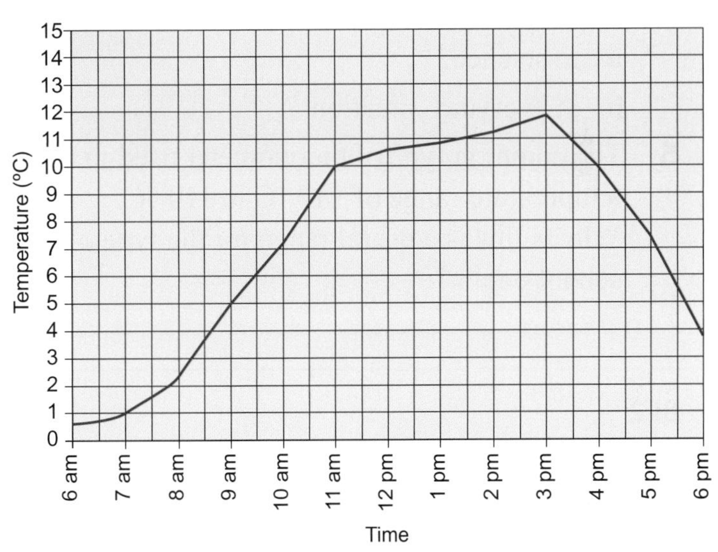

The temperature rose through the day. It reached a peak of about 11 °C at 3 p.m. It started to fall down to about 4 °C by 6 p.m. During the night before, it was only just above freezing because the temperature at 6 a.m. was between 0 °C and 1 °C.

The advantages of using a datalogger in this case were:

- Once the datalogger was set up, the experimenters could go away. They didn't have to stay around in the cold and remember to take readings.
- The three different readings could all be taken at the same time without needing three people, one for each reading.

3

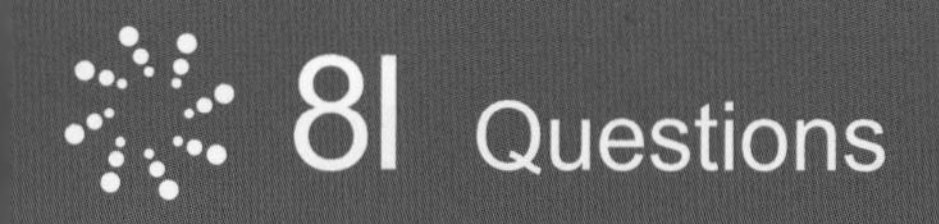

8I Questions

8I.1

1 Some people test how hot or cold a bath for a small child is by dipping an elbow in the water.

a Why might this be a dangerous way of telling the temperature?

b What should they do instead?

Give reasons for your answers.

2 What does the temperature of something tell you?

3 What are the advantages and disadvantages of a liquid-in-glass thermometer compared with a digital thermometer?

4 Which of these temperature scales is used most of the time

a in science?

b in everyday situations?

5 The common type of thermometer used in school laboratories has a temperature range of −10 °C to 110 °C.
Why is this is a useful range for the type of science done by school pupils?

8I.2

1 What is the other name for heat energy?

2 Give two examples of heat moving from a hot place to a cold place.

3 **a** What happens to the particles of a substance when it is heated up?

b What instrument would you use to measure the difference?

4 What is the name for the process by which heat energy passes through a solid?

5 **a** What is meant by a good thermal conductor?

b Give two examples of materials that are good thermal conductors and an example of one in use.

6 **a** What is meant by a good thermal insulator?

b Give two examples of materials that are good thermal insulators and an example of one in use.

7 Give two examples of trapped air being used as an insulator.

8 Why do the handlebars of a metal bike feel colder than the plastic saddle even though they are both at the same temperature?

8I.3

1. Where does convection happen?
2. Describe how a convection current transfers heat energy.
3. Give one example of convection currents in nature.
4. What is the name for the process that transfers heat energy from the Sun to us?
5. Describe what a solar cooker is and how it works.
6. How can thermal radiation be useful after an earthquake?

8I.4

1. Give three ways of keeping heat in a house. Explain why each one works in terms of conduction, convection or radiation.
2. What one advantage does double glazing have, other than keeping heat in?
3. Why are runners wrapped in shiny foil blankets at the end of a long race?
4. Describe how a vacuum flask keeps the heat in, in terms of conduction, convection and radiation.

8I.HSW

1. What are the advantages of using a datalogger to monitor the temperature in the beaker rather than a normal thermometer?
2. Why is a datalogger a useful method for taking readings about the effect of sunlight on the pond?
3. A datalogger is basically a simple computer with one or more sensors attached. It runs on electricity.

   Make a list of possible problems you might have to solve if you used a datalogger to monitor conditions at a pond over a period of 48 hours.

8J.1 About magnets

Scientific enquiry

You should already know

Outcomes

Keywords

A material that produces a force

Some pieces of rock **attract** small pieces of iron or steel.

The rock is called lodestone. The modern name for lodestone is magnetite.

If you hang a piece up or float it on water, one end of it points to the north. It was used like this as a **compass** in ancient times.

One piece of lodestone will attract another piece of lodestone. If you turn one of the pieces round, the pieces of lodestone push each other away. We say that they **repel** each other.

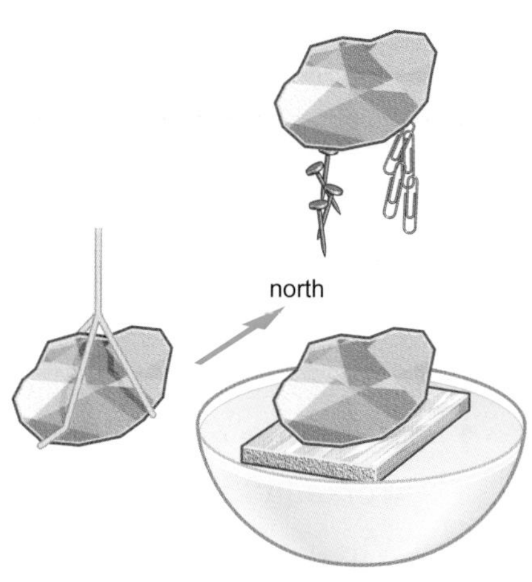

Some rocks are magnetised. They attract things made of iron or steel.

Question 1 2

If you rub a strip of iron with lodestone, the iron will behave like the lodestone. Rubbing with a lodestone makes iron into a magnet.

You can divide substances into three types.

Magnets	Magnets attract or repel other magnets depending on which way they point. They also attract magnetic materials. They are often made from iron.
Magnetic materials	Magnetic materials are attracted to magnets but do not attract or repel each other.
Non-magnetic materials	These materials are not attracted to magnets.

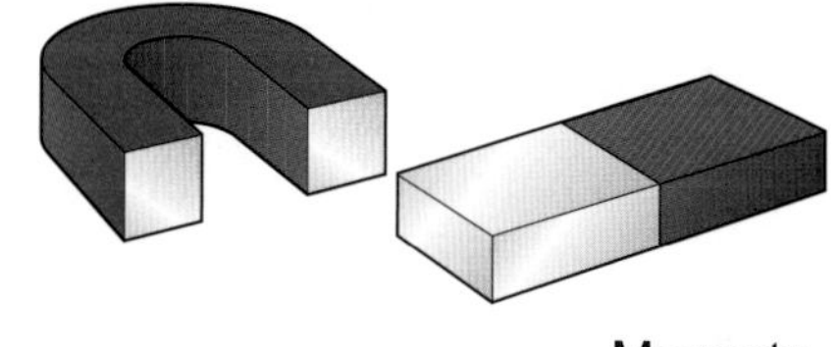

Magnets.

Most people think that metals are magnetic materials. This is wrong. Of the eighty or so elements that are metals, only three are attracted to a magnet at normal room temperature. These are **cobalt**, **nickel** and **iron**.

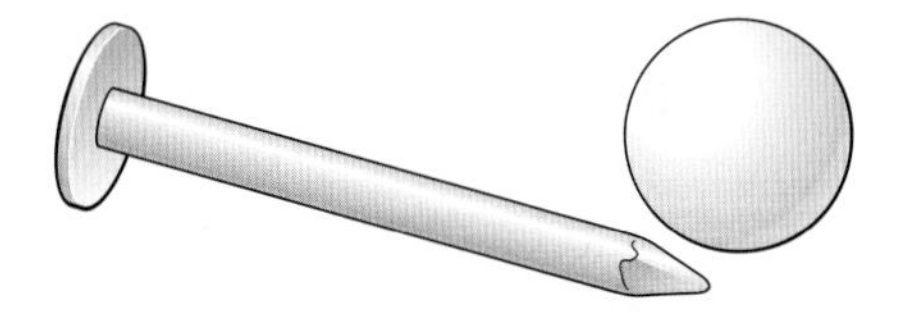

Magnetic materials.

Steel is a material that contains iron. There are different types of steel. Some are magnetic and some are not – like high-quality stainless steel.

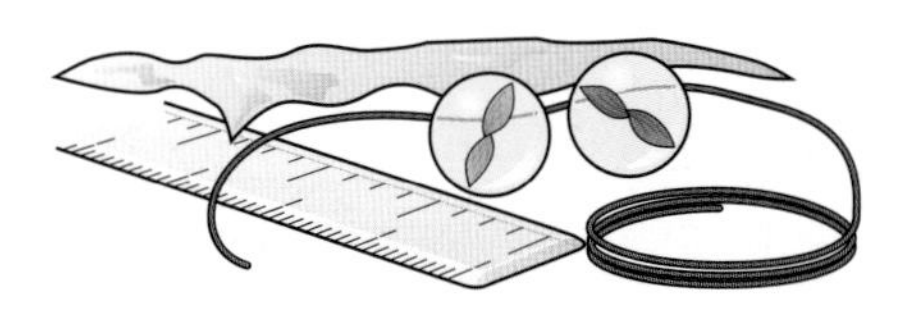

Non-magnetic materials.

Question 3 

Magnetic forces

Objects held on a fridge door by magnets.

The magnetic force will pass through non-magnetic materials like paper, plastic, paint, skin and bone.

The force between a magnet and another magnet or some magnetic material works through many other materials. We use this idea to keep fridge doors closed.

We also use magnets to hold notes on fridge doors. The small magnets sold to do this are often called 'fridge magnets'.

When a magnet can move freely and it is not near other magnets or magnetic materials, it always rests with one end pointing towards the north of the Earth. This end of the magnet is called the **north-seeking pole** or north pole for short. The other end points towards the south. This is called the **south-seeking pole** or south pole for short.

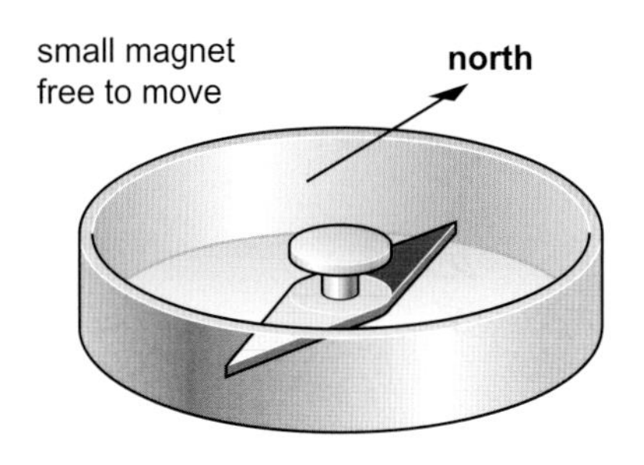

The forces produced by a magnet are strongest at its poles.

The only thing that will repel a magnet is another magnet.

If you bring magnets together to see which combination of poles attract and which repel, you get these results.

Poles	Force between poles
north pole near another north pole	repulsion (= pushing apart)
south pole near another south pole	repulsion (= pushing apart)
north pole near a south pole	attraction (= pulling together)

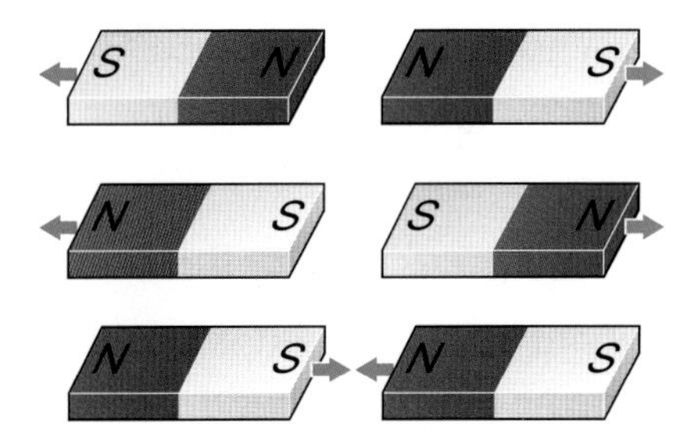

These results produce a basic law.

- Two poles that are different attract each other.
- Two poles that are the same repel each other.

With a compass, you can use this law to find out which part of a magnet has which pole.

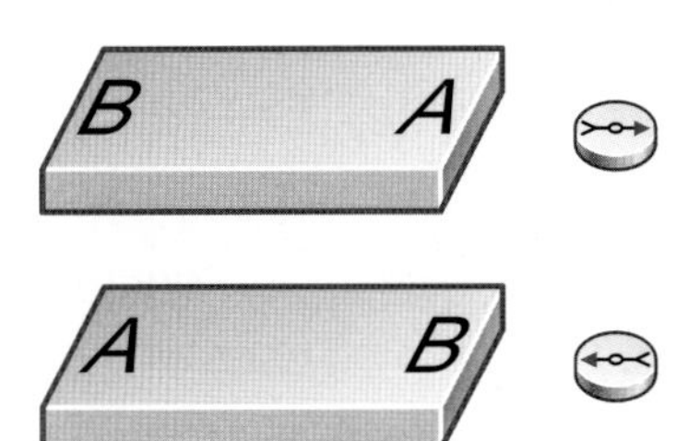

End A attracts the south pole of the compass and repels the north pole. End B attracts the north pole of the compass and repels the south pole. So, end A is a north pole and end B is a south pole.

8J.2 Magnetic fields

You should already know | Outcomes | Keywords

What is a magnetic field?

In science, the term **magnetic field** is used to name an area in which you can detect magnetic forces.

The magnetic field is where a magnet will attract magnetic materials or push or pull on another magnet.

You can get a picture of a magnetic field with iron filings or by using small compasses. The diagrams show you how.

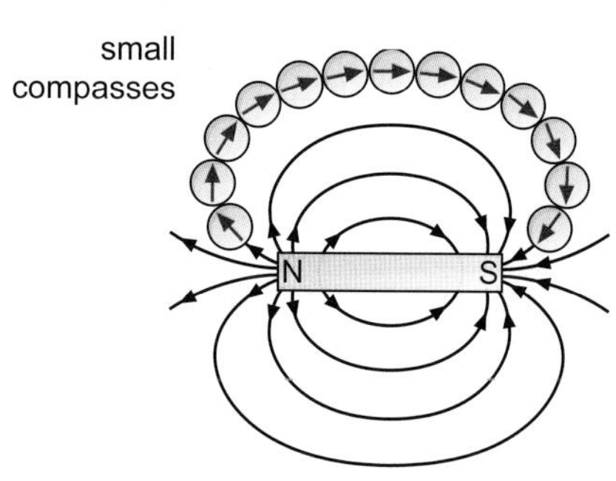

The black lines with arrows show how the compasses point when they are placed there.

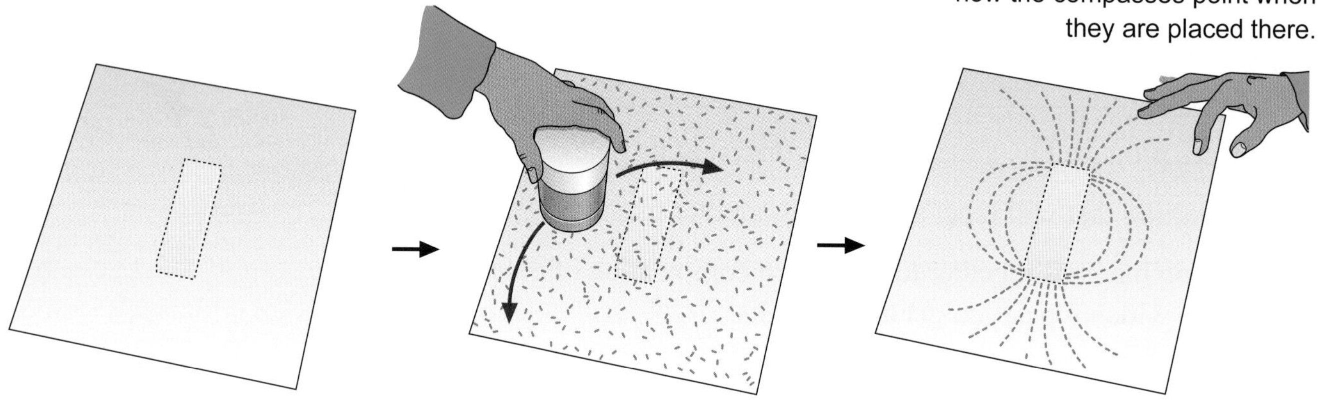

Put a piece of card on top of the magnet.

Sprinkle iron filings as evenly as you can.

Tap the card with your fingers.

What do the lines in the magnetic field show?

We draw lines round magnets called lines of magnetic force or **magnetic field lines**. The lines show the direction in which a small compass will point if you put it on the line at that point.

The small arrows on the lines show the direction the north pole of the compass points in.

Magnets are strongest at their poles. The lines get closer at the poles.

Where lines are close together on a map of a magnetic field, the force of magnetism is stronger.

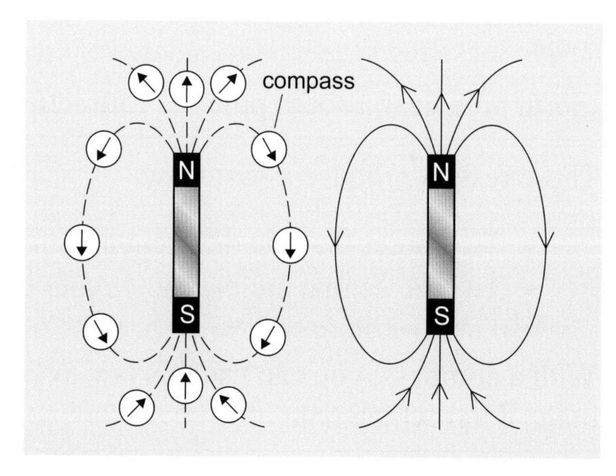

There is a magnetic field all around a magnet.

We can show a magnetic field using lines of magnetic force.

Question 1  

The Earth's magnetic field

The needle of a magnetic compass is a small magnet. This magnet is free to turn.

The needle of a magnetic compass rests with one end pointing north and one end pointing south. This happens because the Earth has a magnetic field. The lines of force in the Earth's magnetic field run from the south of the Earth to the north of the Earth.

In 1600, a scientist called William Gilbert wrote the first book about magnets. In it, Gilbert suggested that the Earth's magnetic field is like the field of a giant bar magnet inside the Earth.

The puzzling thing is that, if there was a large magnet inside the Earth, the pole under the geographic north pole would be a magnetic south pole!

magnetic north pole

geographic north pole

The imaginary magnet that could produce the Earth's magnetic field lines. Its south pole points towards the Arctic and its north pole points towards the Antarctic.

A compass points to a place we call **magnetic north**. This is near the north pole of the Earth. Its position is always changing, very slowly. The table shows how it has moved in the past 400 years.

It moves because some of the Earth's core is a molten liquid. This has iron compounds moving about in it. These are the main cause of the Earth's magnetic field.

Year	Position of magnetic north
1580	11° east of geographical north
1700	7° west of geographical north
1800	24° west of geographical north
1900	17° west of geographical north
1960	8° west of geographical north

Question 4 | 5

Making a magnet

You can use the field from a magnet to make a piece of iron into a magnet.

An easy way to do this is to stroke a piece of iron with one pole of a magnet. The diagram shows you how.

As the piece of iron is stroked many times in the same direction, the magnetic field from the magnet causes the particles inside the iron to line up so that they produce a magnetic field of their own.

In the diagram, the end of the iron on the right-hand side will become a south pole.

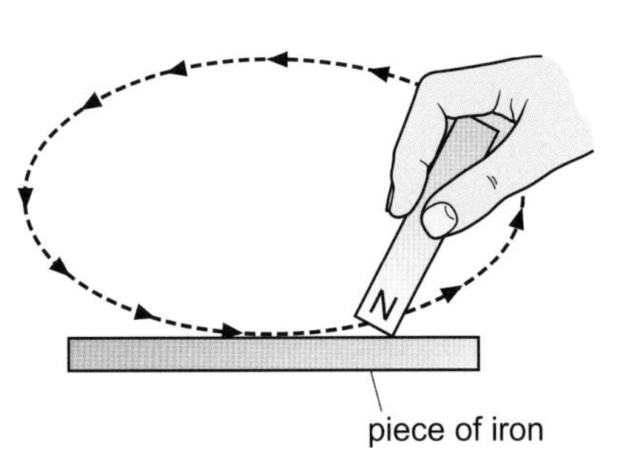

Check your progress

8J.3 Electromagnets

You should already know | Outcomes | Keywords

Electricity and magnetism

An electric current has a magnetic field around it.

This was discovered by accident by a scientist called Hans Christian Oersted in 1819.

He noticed that a compass changed direction when he turned on an electric current nearby.

You can show the effect easily with a coil of wire. The diagram shows how.

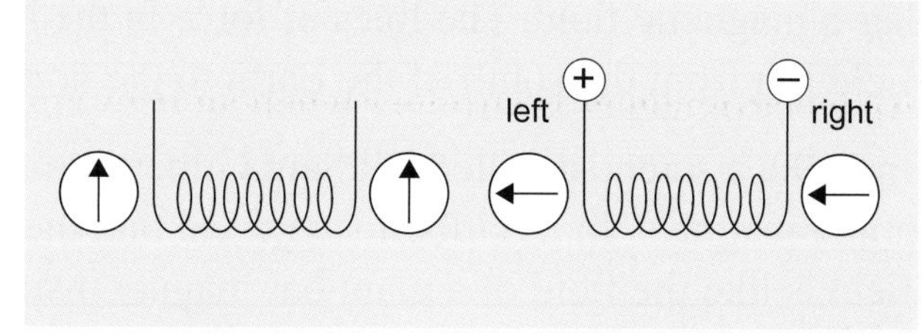

A coil of copper wire does not move a compass needle. The coil is not a magnet.

When a current flows through the coil, the compass needle moves. The coil is just like a magnet.

A coil of wire connected to a suitable voltage supply has a magnet field around it when the current flows. It is called an **electromagnet**.

Question 1 | 2

On and off

You can turn an electromagnet on and off. You do this by turning the current on and off. You cannot do this with a permanent magnet.

If you use a permanent magnet to pick up bits of iron and steel, you have to pull the bits off.

With an electromagnet, you can just use the switch.

This idea is used in scrap yards to move wrecked cars.

A bar magnet stays magnetised all the time. We call it a permanent magnet.

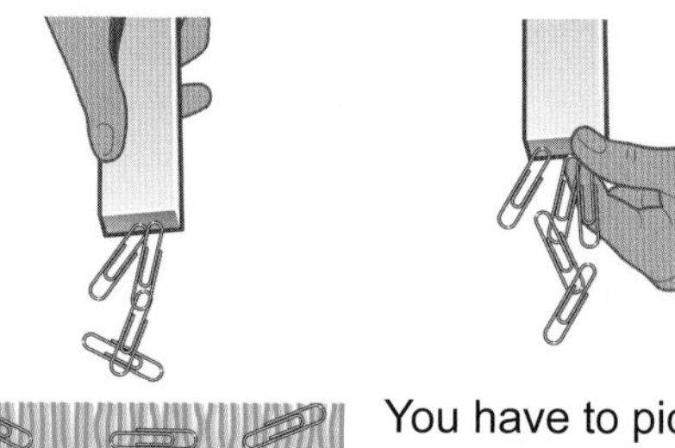

You have to pick the paperclips off.

The electromagnet in the crane lifts a scrap car.

When the crane driver switches off the current, the car falls.

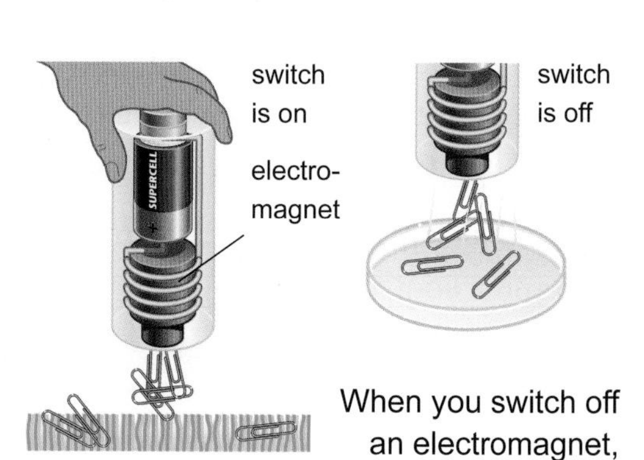

When you switch off an electromagnet, the paperclips fall off.

Question 3 | 4

How do electromagnets work?

Wires carrying an electric current produce a magnetic field.

The field pattern round a coil of wire is similar to the one around a bar magnet.

If a piece of iron is then placed inside the coil, the iron is magnetised. A piece of iron used in this way is called a core. An electromagnet is stronger when an **iron core** is used.

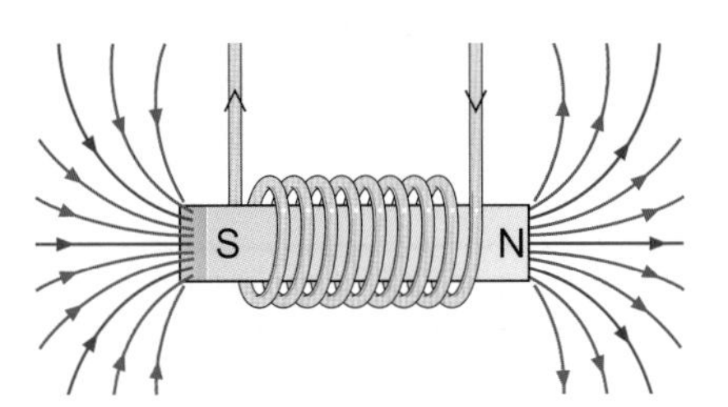

The magnetic field of the iron core adds itself to the magnetic field of the coil.

It is important to use 'soft iron' for the core of an electromagnet. Soft iron does not stay magnetic when the electromagnet is switched off.

If you use a piece of steel for the core, then it will still be a magnet when the current is switched off.

You can use the coil of an electromagnet to make a permanent magnet from a piece of steel.

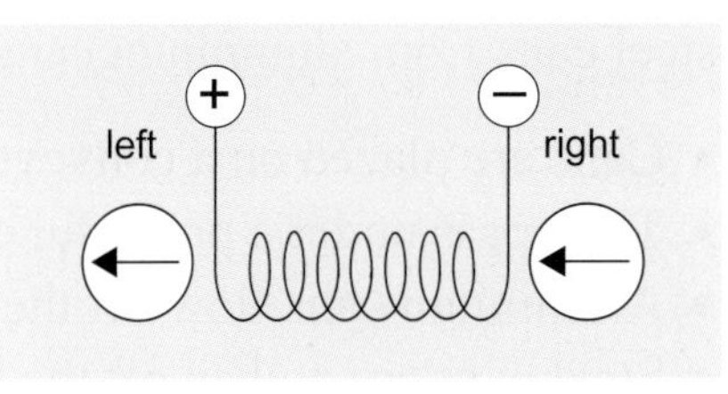

When a current flows through the coil, the compass needle moves. The coil is just like a magnet.

Reversing the poles

If you change the way the current flows in an electromagnet, you change the poles.

The diagrams show what happens when you change over the connections to the power supply.

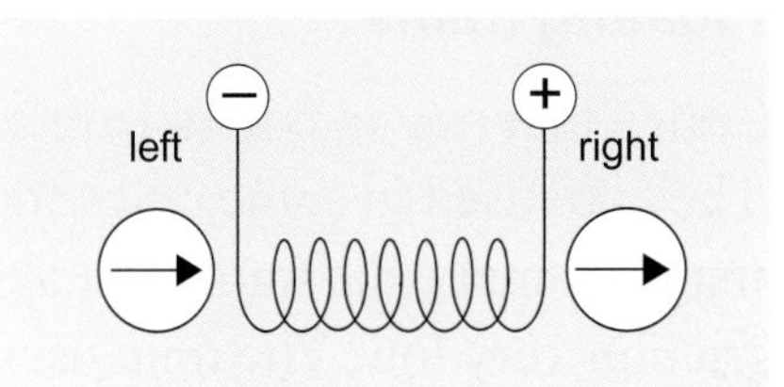

This is what happens when the current in the coil is reversed.

Changing the strength of an electromagnet

You can change the strength of an electromagnet in three ways.

Way of increasing the strength	Notes
increase the current	Larger electric currents produce stronger magnetic fields. The wires must be thick enough to carry the high current without melting.
have more turns	The more turns you have on an electromagnet, the stronger it is. The problem is that the size of the magnet also increases.
put some soft iron inside	A piece of soft iron inside a coil is called a core. It makes the field stronger. Soft iron is used because it loses magnetism when the current is switched off.

You should already know

Outcomes

Keywords

Separating cans

The steel used to make food cans is a magnetic material. The aluminium in drink cans is not a magnetic material.

An electromagnet can be used to separate a mixture of steel cans from aluminium cans in a recycling plant.

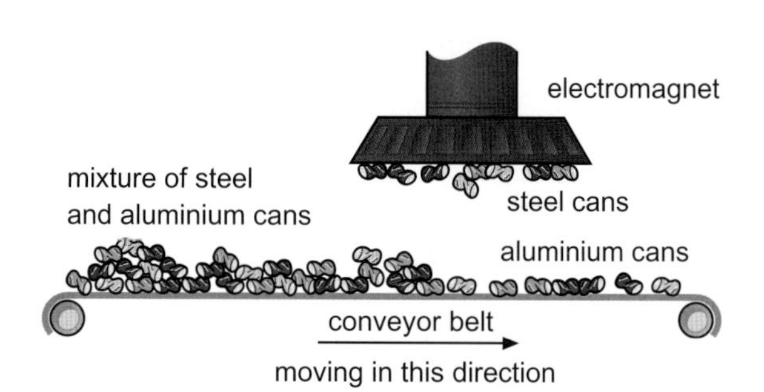

- Cans are placed on a conveyer belt.
- They pass under a powerful electromagnet.
- Aluminium cans flow off the end of the conveyer belt.
- Steel cans are picked off the belt by the electromagnet.

Question 1

Floating trains

Trains that run on one rail are called monorail trains. They are used in some cities for public transport. In one design, magnets make the train float above the rail. This makes the friction very low. The train uses less energy than a train running on wheels.

The diagram shows a design in which the magnets are attracted up to the rail. This force actually lifts the train off the rail.

levitating magnets

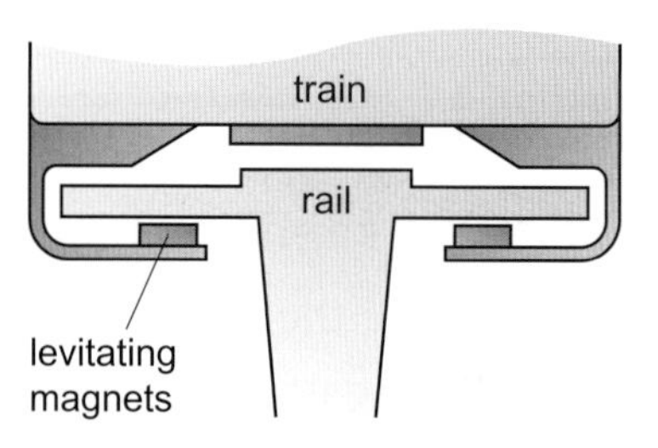

Question 2

Recording information

Magnetic tape can be used to store information. The tape contains tiny magnetic particles. The information is stored by magnetising the particles in different patterns.

This idea is used in some car parks to check tickets. The magnetic strip on the ticket records the time you arrive at the car park.

The pay machine reads your arrival time from the strip and tells you what to pay. When you pay, the machine records it on the strip so that the ticket will raise the barrier on the way out. If you put your ticket in a bag with a magnetic catch, the ticket might fail to work!

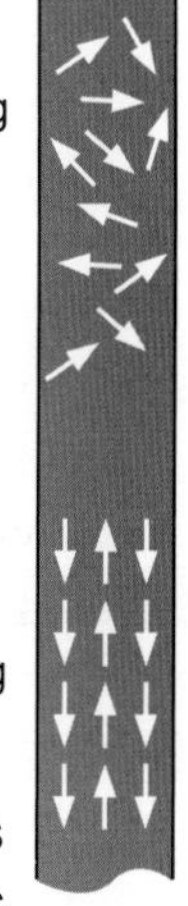

When the information is recorded on the magnetic strip, the magnetic particles line up into a pattern.

Question 3

Bells

The diagram shows an electric bell circuit that uses an electromagnet. The iron core with the coils around it is an electromagnet.

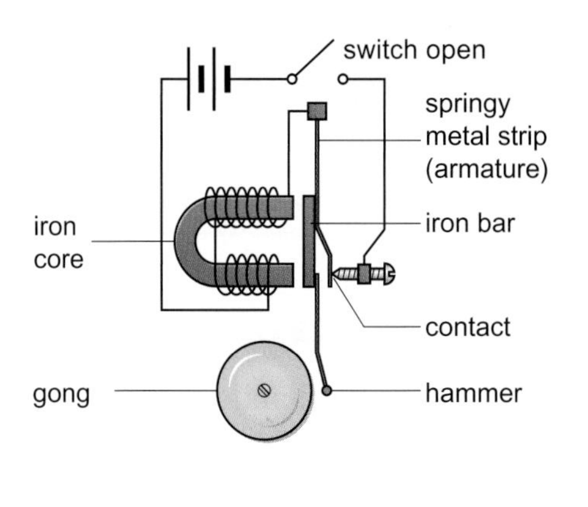

When the switch is pressed, a cycle of events take place.

- The current flows round the circuit.
- The electromagnet is turned on.
- The iron bar attached to the hammer is attracted to the iron core of the electromagnet and moves.
- The hammer hits the gong.
- When the iron bar moves to the left, the contact on its right is broken.
- The current stops flowing and the electromagnet turns off.
- The iron bar is no longer attracted to the electromagnet and the spring moves it back.
- The contact on the right of the iron bar is made.
- The circuit is complete, so current flows around the circuit and the cycle starts again.

Question 4

Relays

The **relay** is a very important use of an electromagnet. A relay is a switch that uses an electromagnet, so a current in one circuit turns on a current in another.

The diagram shows how a pressure switch will turn on an electromagnet in a relay. The iron in the relay moves and makes the contact in the second circuit to light the bulb.

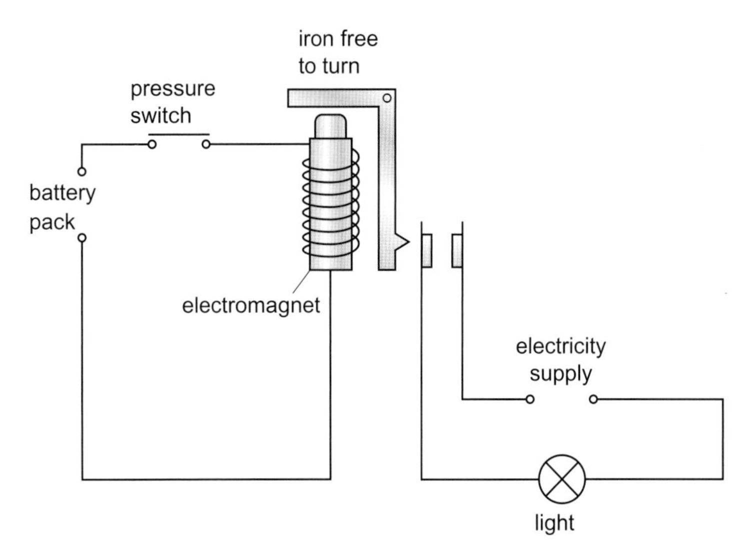

This circuit controls a light that switches on if someone steps on the pressure switch.

Car starter motors use relays. A small current flows in one circuit when the ignition key is turned. This makes a relay switch on a very large current to the starter motor.

The relay is a way for a small, safe current to switch on a large, dangerous one.

Question 5

Review your work

Summary ➡

8J.HSW Different types of data

You should already know

Outcomes

Keywords

Gathering data

Scientists record and analyse data.

Sometimes they get data from their own experiments. These are examples of **primary sources**

Sometimes they get data from other people or the Internet. These are called **secondary sources**

Wherever the data comes from, it can be classed as either **qualitative** or **quantitative**

Remember

Scientists use data to provide evidence for scientific explanations.

Qualitative	the data is in the form of words or pictures
Quantitative	the data is in the form of numbers

The diagram shows an investigation into how strong two magnets are. The magnet that can hold the most paper clips in a chain will be the strongest.

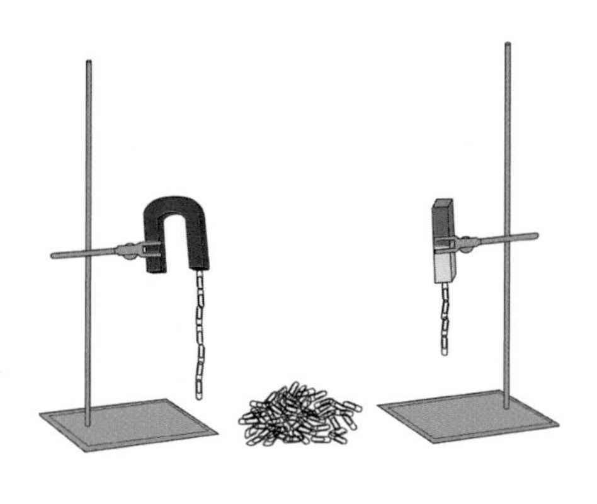

This is the data from this experiment.

Magnet	Strength measurement
bar magnet	four paper clips
horseshoe magnet	seven paper clips

These strength measurements are examples of quantitative data.

If you tested some different materials to see if they were magnetic, you might get results like this.

Object	Observation
plastic ruler	not attracted to a magnet – not magnetic
marble	not attracted to a magnet – not magnetic
iron nail	attracted to a magnet – magnetic

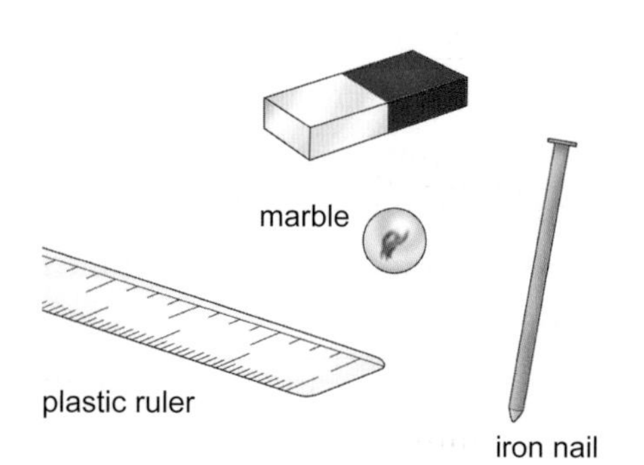

These are examples of qualitative data.

Investigations with magnets

The diagram shows an investigation to find out if the magnetic force will pass through different thicknesses of paper.

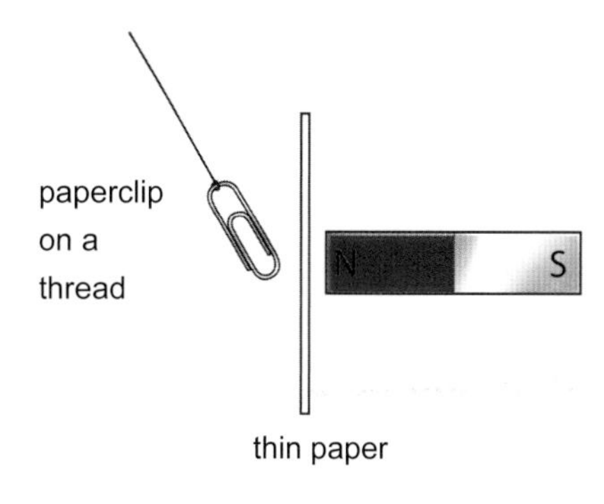

Different thicknesses of paper are placed between the magnet and the clip. The angle between the thread and the vertical is a measurement of the magnetic force.

A pupil who did the investigation produced these results.

Thickness of paper	Angle of thread from vertical
thin	31°
thicker	31°
very thick	31°

Question 3

Investigating the strength of an electromagnet

A similar idea can be used to investigate the strength of an electromagnet. The electromagnet attracts the steel nut. You use the angle the thread makes with the vertical to measure how strong the magnet is.

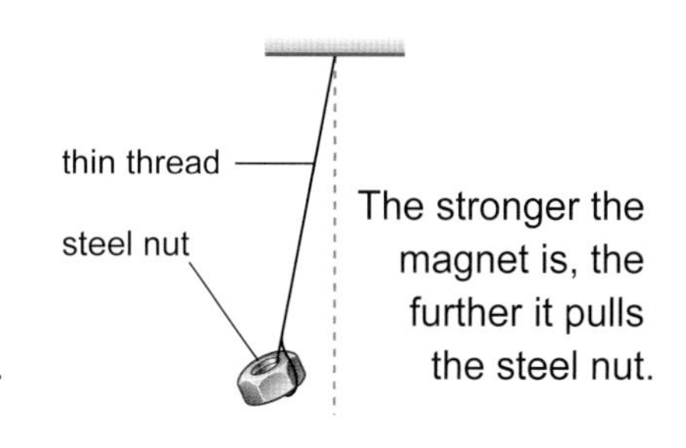

There are three ways of changing the strength of an electromagnet. So you need three investigations to test the strength. The diagrams show part of the experimental set-up for each investigation.

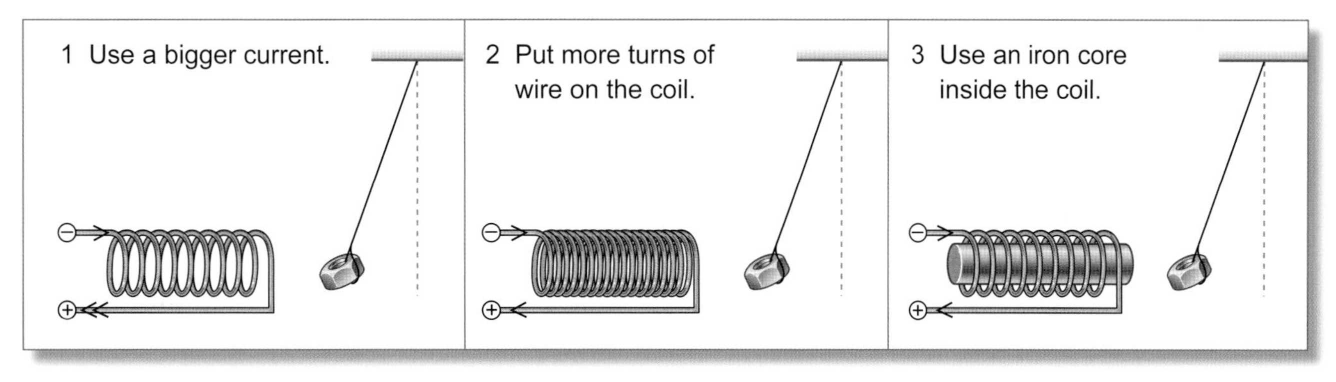

To get quantitative data, you would also need:

- an ammeter to measure the size of the current flowing in the electromagnet
- a protractor and plumb line to measure the angle between the thin thread and the vertical.

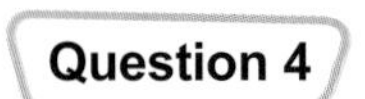

8J.1

1 What is lodestone?

2 What is a compass and what does it do?

3 What is the difference in behaviour between magnets and magnetic materials?

4 What is wrong with this statement?

'Steel is attracted to a magnet.'

5 a Why are magnets in plastic strips used to seal fridge doors?

b What does this mean for the material used for the casing of a fridge?

6 Describe how different combinations of magnetic poles attract and repel each other.

7 How can you use a compass to decide if a piece of metal is a magnet or not?

8J.2

1 What is the name for an area where you can detect magnetic forces?

2 Describe how you can use iron filings to get a picture of the magnetic field around a magnet.

3 What do the lines of magnetic force show, and how can you tell where the field is the strongest?

4 Why does a compass point to the north of the Earth?

5 Why does the Earth have a magnetic field?

6 a Describe how you can make a magnet.

b How could you tell which pole was which on the magnet you make?

8J.3

1 What is there around a wire carrying an electric current? How can you tell?

2 What is an <u>electromagnet</u>?

3 Give <u>two</u> advantages of an electromagnet over a magnet made from a piece of metal.

continued

4 a Why is an electromagnet useful on a crane in a scrap yard?
  b Is it useful for every possible type of car?
  Give a reason for your answer.
5 What happens if you reverse the current in an electromagnet?
6 Describe how the strength of an electromagnet can be changed.

8J.4

1 An electromagnet is used to separate steel cans from aluminium cans.
  Which type of can attaches to the electromagnet?
  Give a reason for your answer.
2 What do the magnets do in a monorail train?
3 Why is it a bad idea to put a parking ticket with a magnetic strip in a bag with a magnetic catch?
4 In an electric bell, which part is attracted to the electromagnet?
5 a What does a relay do?
  b Explain how an electromagnet operates in a relay.

8J.HSW

1 What is the difference between secondary sources and primary sources?
  Give one example of each.
2 What is the difference between quantitative data and qualitative data?
  Give one example of each type.
3 What are the different types of data in the pupil's experiment to test whether the magnetic force passes through paper?
4 For each experiment shown in the diagram, describe what you would change (the independent variable), what you would measure (the dependent variable) and what you would keep the same (control variables).
  For each variable, state whether it is quantitative or qualitative.
5 Describe how you would carry out one of the three experiments, including how you would use the protractor and plumb line to measure the angle of the thread.

Scientific enquiry

8K.1 Travelling light

You should already know | Outcomes | Keywords

Where does light come from?

A **light source** is anything that produces light. The Sun is our main light source. But, since prehistoric times, humans have used other light sources to help them see when it is dark.

Electric light sources can be switched on and off instantly. Electricity is safer and more convenient than candles and oil lamps.

Electric lights do not have flames in them. There is little risk of starting a fire.

Some light sources.

Light travels in straight lines

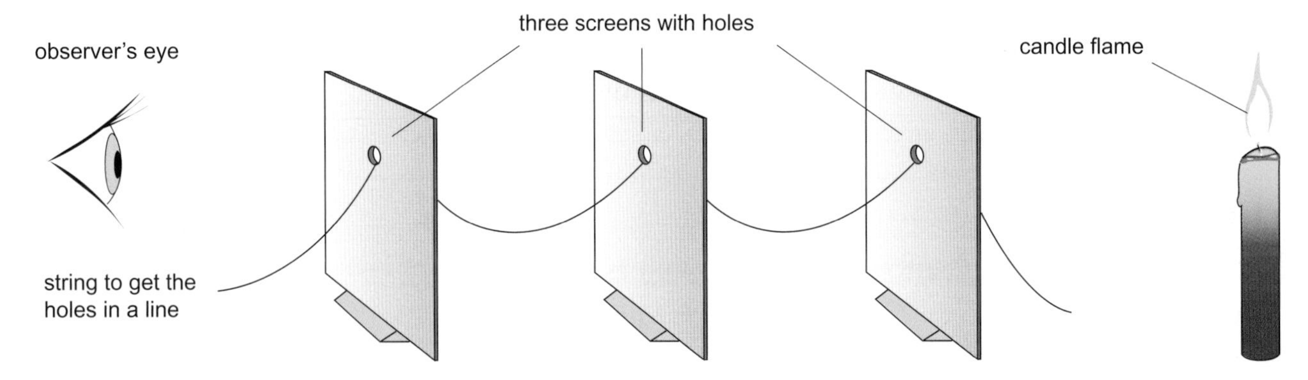

We use straight lines with arrows to represent light. We call each line a **ray**. A light ray shows the path that the light follows. A diagram with rays to show what light does is called a ray diagram.

The ray diagram shows how a torch and a pencil can be used to make a shadow. The light cannot go through the pencil. Because light travels in straight lines, it cannot go around the pencil. That is why the pencil casts a shadow when it is put in front of the torch.

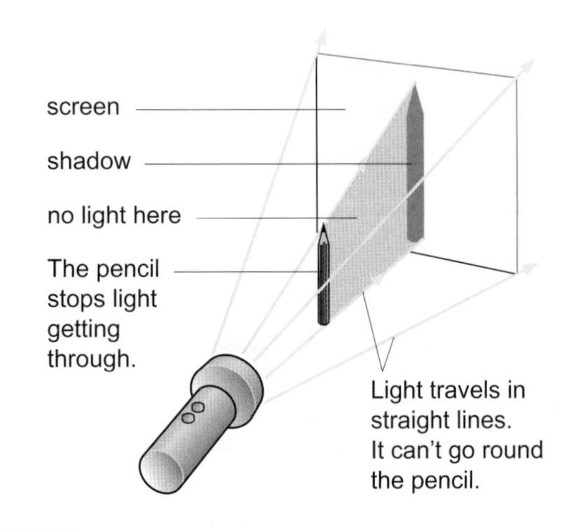

Question 1 | 2

The speed of light

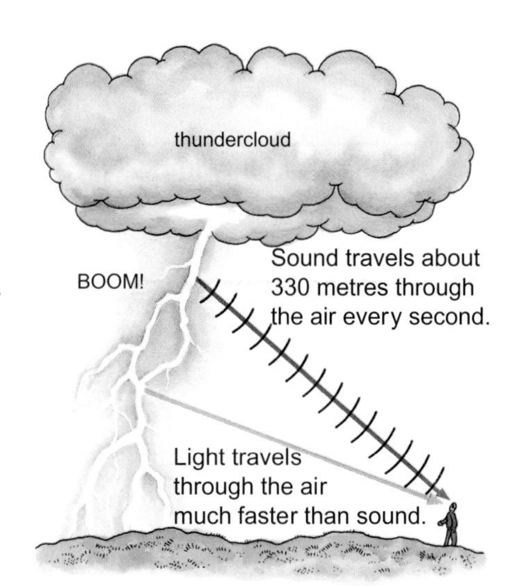

If you go to watch a cricket match, you see the ball being hit before you hear it. Light travels at an incredibly high speed. It seems to reach you straight away. Sound travels about a million times slower than light.

You can use this effect to estimate how far away a thunderstorm is.

Thunder is the sound produced by a bolt of lightning. If a thunderstorm is approaching in the distance, you see the lightning a long time before you hear the thunder.

The speed of light is so high that you see the lightning instantly. The sound takes about three seconds to travel a kilometre.

The distance to the storm is about one kilometre for every three seconds you can count between seeing the lightning and hearing the thunder.

Question 3

What is the speed of light?

The first attempt to measure how fast light travels was by the Italian scientist Galileo Galilei in 1600. He tried uncovering a lantern and timing how long the light took to travel a few miles to another person. His idea did not work. The speed of light was too fast to measure by his method.

The first accurate measurements of the speed of light were made about 150 years ago using a spinning mirror. We now know that light travels 300 000 000 metres in one second.

It is almost impossible to imagine what a speed of 300 000 000 m/s means.

Distance	Time for light to travel that distance
from a lamp to the door of a room (about 3 m)	one hundred-millionth of a second
from a lighthouse to a ship 30 km away	one ten-thousandth of a second
from the Sun to the Earth (150 000 000 km)	about 500 seconds (8 minutes and 20 seconds)
from the nearest star to us (about 40 400 000 000 000 km)	about 135 000 000 seconds (about 4.25 years)
from the edge of the observable universe to us (about 100 000 000 000 000 000 000 000 km)	about 10 000 million years

Question 4

8K.2 Light hitting objects

You should already know

Outcomes

Keywords

What happens when light hits an object?

When light hits something, it can do one of three things.

- It can go through (be **transmitted**).
- It can bounce back (be **reflected**).
- It can stay inside and heat up the object (be **absorbed**).

Transparent substances

Some substances let almost all of the light go straight through them. Glass, water, air and some types of plastic do this. We say these substances are **transparent**.

If you can see clearly through the substance, we say that it is transparent.

Question 1

Translucent substances

Sometimes you want to let the light through but you do not want anyone to be able to see through.

Substances that let light through but do not let you see a clear image through them are called **translucent** substances. Examples of translucent substances in use are

- frosted glass in a bathroom window
- sheets of white cotton used as sun blinds.

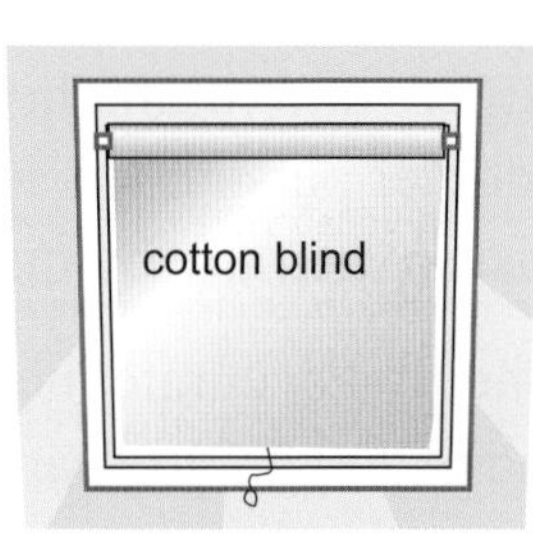

Some light gets through, but you can't see through the frosted window or the blind.

Translucent substances scatter light in all directions as it passes through, so you cannot see a clear image through them.

Clouds are translucent. The Sun's light is scattered when it comes through them. That is why you do not get shadows on a cloudy day even though it is still light.

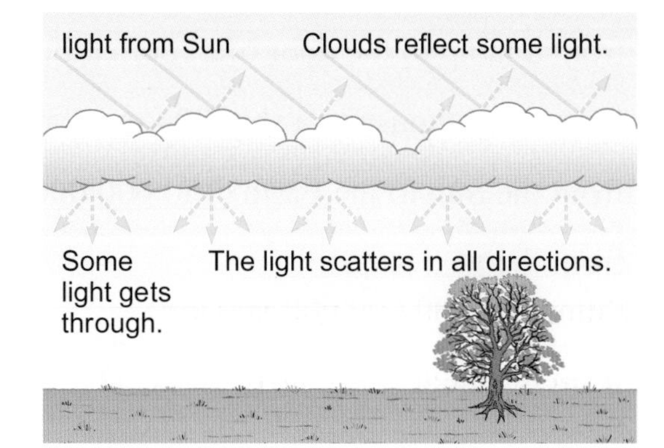

Question 2 3

Opaque substances

Some substances stop light going through them. A substance that stops light is **opaque**. Metal and wood are opaque substances. Some types of plastic are also opaque.

Opaque substances are used for blackout blinds. They are used in photography to stop light reaching light-sensitive film.

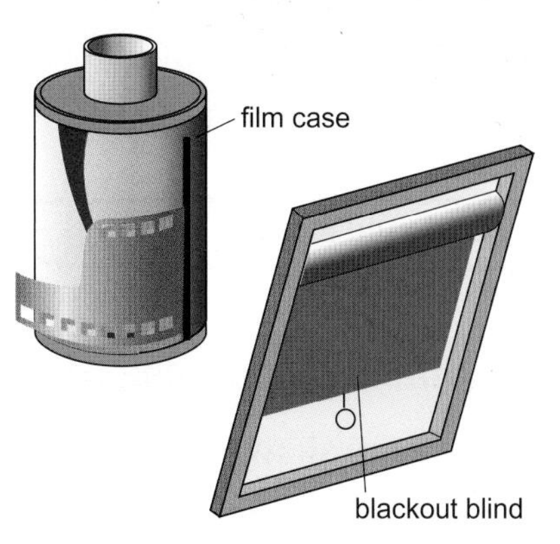

You must use an opaque substance for the case of a photographic film or the light will spoil it. The windows in a roof can be fitted with opaque blinds to shut sunlight out completely.

Question 4

What happens when light hits a surface?

When light hits a surface, some of it bounces off. We say that it is reflected.

- Transparent glass only reflects a very tiny amount.
- Opaque substances can reflect almost all of it. How much light is reflected depends on the colour of the surface.
  - Pale or white things reflect most of the light that falls on them. That is why they look pale or white.
  - Black and dark things absorb most of the light that falls on them. That is why they look black or dark.

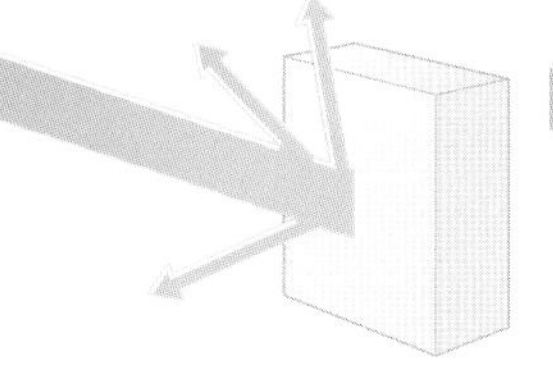

White things and pale things reflect most of the light that falls on them.

Black things and dark things reflect very little of the light that falls on them. They absorb most of the light.

You see an object because the light from it enters your eye. Light is given out by objects like a light bulb, and travels in a straight line from it to your eye.

For objects that do not give out light, you see them when light reflects off them into your eye.

Some things give out their own light. We can see other things because they reflect light.

The only luminous object in this room is the light bulb. When it is not on, none of the objects in the room can be seen.

Objects that give out light are called **luminous** objects.

Objects that do not give out light are called **non-luminous** objects.

Question 5 **6**

8K.3 Mirrors

You should already know | Outcomes | Keywords

How different surfaces reflect light

The diagram shows a lamp shining on a picture. The picture has a surface that has a lot of very tiny bumps on it. The bumps make the light reflect in all directions.

A shiny surface is one that is very smooth. When light reflects off a shiny surface, it is not scattered in all directions. You can see a clear reflection in a shiny surface.

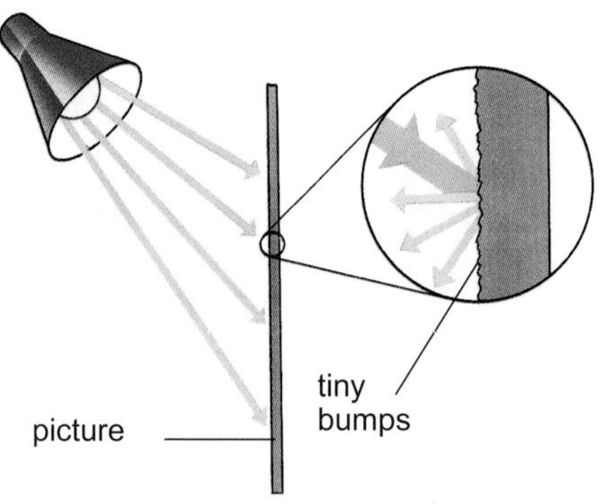

The picture has a bumpy surface. So it reflects light in all directions.

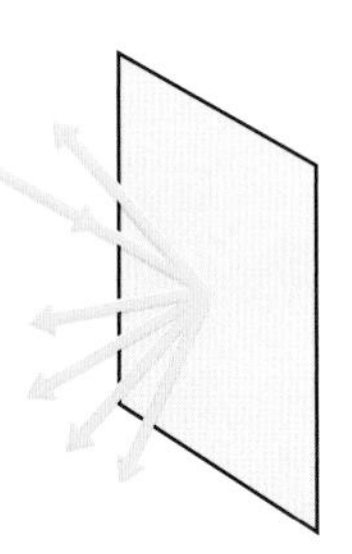

A piece of paper scatters light in all directions.

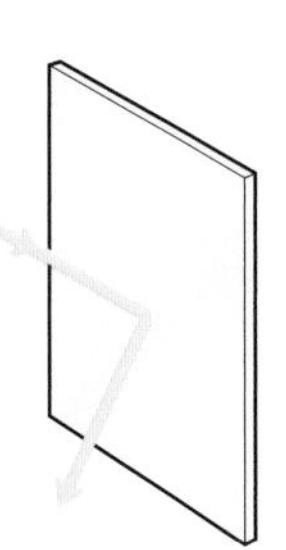

A mirror reflects all the light in the same direction.

How a mirror reflects light

A mirror reflects light so that it bounces off at the same angle as it hits the mirror.

We measure the angle of the light hitting and leaving the mirror from a reference line called a **normal**. The normal is a line drawn at 90° to the surface of the mirror.

The ray of light hitting the mirror is called the incident ray. The angle between it and the normal is called the **angle of incidence**.

The ray of light reflecting off the mirror is called the reflected ray. The angle between it and the normal is called the **angle of reflection**.

When light reflects off a mirror, the angle of incidence equals the angle of reflection.

You can see your own reflection in these.

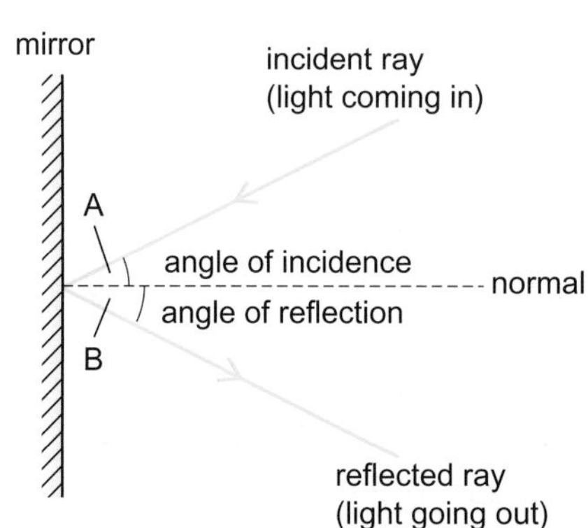

Angle A and angle B are equal. The mirror reflects a beam at the same angle as it strikes the mirror.

Question 1 | 2

Looking in a mirror

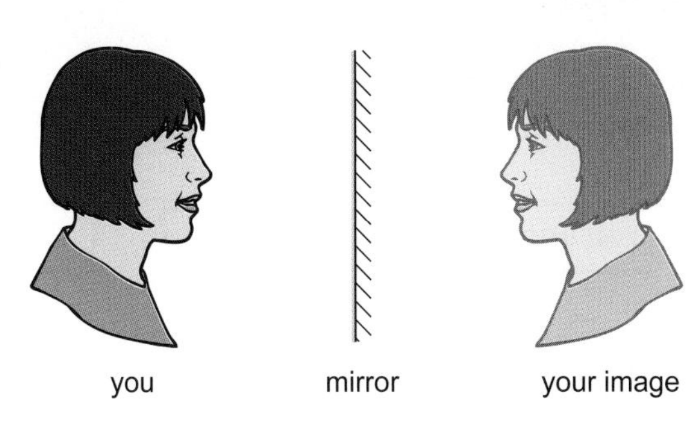

A flat mirror is also known as a plane mirror. When you look into a plane mirror, you see an **image** of your face. We call the thing that the light comes from the **object**. In the diagram shown here, the face is the object.

Because of the way light reflects off a flat mirror, the image of your face follows certain rules.

- The image is the same size as your face. Your face does not look bigger or smaller.
- Your reflection looks as if it is as far into the mirror as your face is in front of the mirror.
- The image is the opposite way around to the object.

Using mirrors

If you want to see over the top of something, you can use two mirrors in an instrument called a **periscope**.

A periscope is made from two mirrors in a tube.

The light enters the periscope, reflects off the first mirror at the top and travels down to the bottom mirror. Then the light is reflected into your eye.

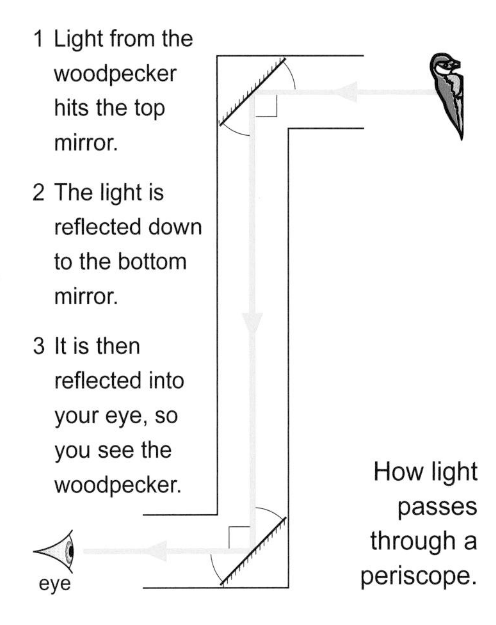

How light passes through a periscope.

Mirrors are also used at dangerous junctions and bends to help drivers see if another car is coming.

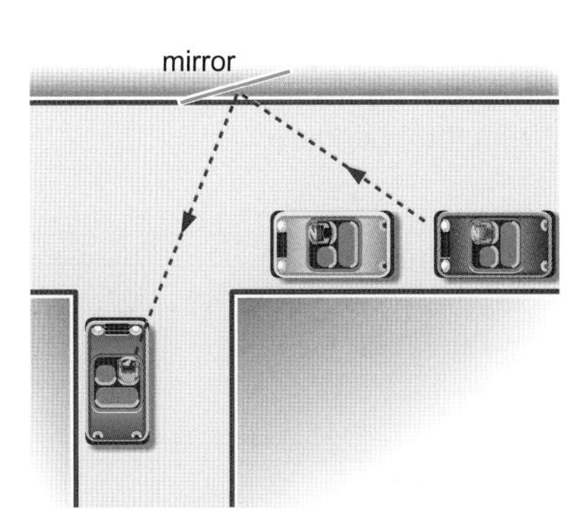

In the diagram, the driver of the blue car can just see the front of the yellow car along a direct line of sight. The driver can see the red car in the mirror.

Because the light takes the same path in the opposite direction, the driver of the red car can see that there is a blue car waiting at the junction.

Question 4

Check your progress

8K.4 Bending light

You should already know | Outcomes | Keywords

Changing direction

You can change the direction of a ray of light by bouncing it off a mirror.

You can also change its direction by shining it into a different transparent substance. When you shine a ray of light at an angle from one substance into another, it changes its direction.

(The dotted line drawn on the diagrams is called the normal.)

When light bends like this, it is called **refraction**. We say that the light has been refracted.

- When there is a large angle between the light ray and the normal, the refraction (bending) is quite large.
- When a light ray travels along the normal, there is no refraction.
- Refraction works in both directions.
- If you shine a ray of light into a transparent substance like water or glass from the air, it bends towards the normal.
- If you shine a ray of light out of a transparent substance like water or glass into air, it bends away from the normal.
- If you shine it along the normal in any direction, it doesn't bend at all.

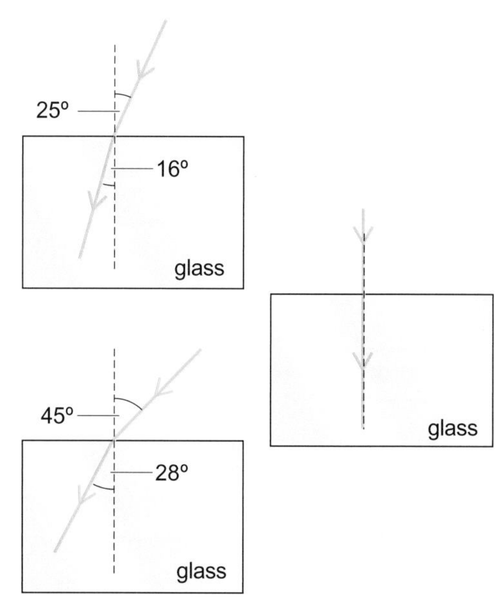

Light does not bend when it goes into glass along the normal.

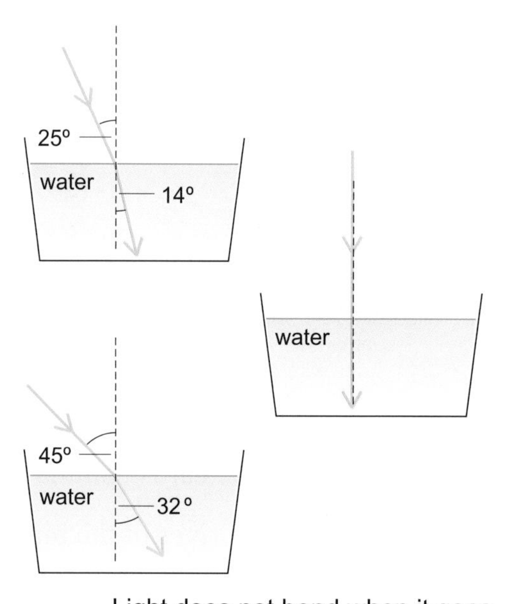

Light does not bend when it goes into water along the normal.

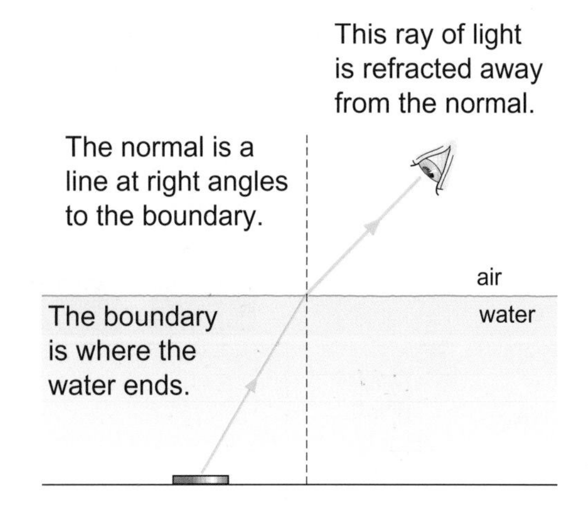

Question 1 | 2 | 3 | 4

Seeing around corners

You can use refraction to see around corners. In the top diagram, Kris cannot see the coin because the light ray that travels past the edge of the metal can does not enter his eye.

In the second diagram, his friend Sam has added some water while Kris keeps his head still.

The coin comes into view because the light ray from it is refracted (bent) as it comes out of the water.

Kris can see the coin now because the light from it enters his eye.

Shallow water

When you look into water, refraction makes it look shallower than it is. This diagram shows what happens. The image of the coin does not appear to be as deep as the coin actually is. It seems to be closer to the surface.

This effect can be dangerous for people who cannot swim and who do not know that water seems shallower than it really is.

The same effect can be seen in other situations.

- A straight stick looks bent when you put it into water.
- A swimmer looks shorter stood under the water in a pool.
- A fish sees a fly in a different position than it actually is.

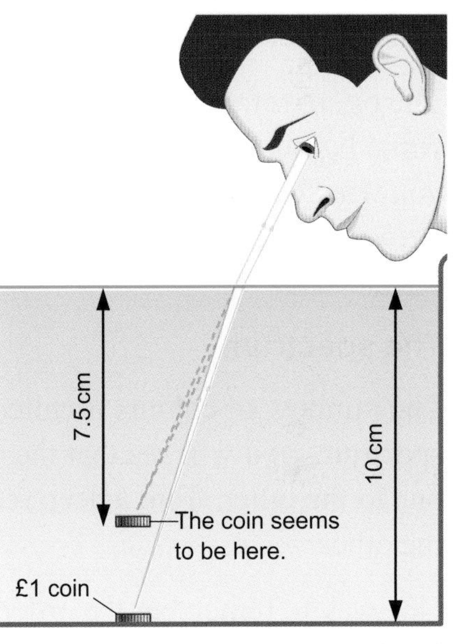

The coin looks closer than it actually is.

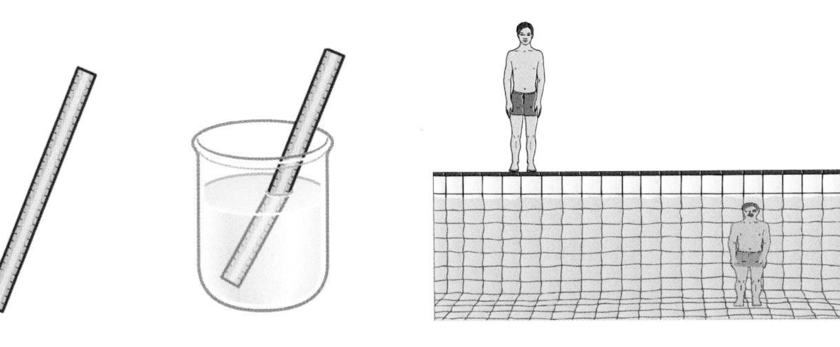

This ruler is straight.

But if you dip it into water, it looks bent.

The boy looks shorter in water because light is refracted as it goes from water to air.

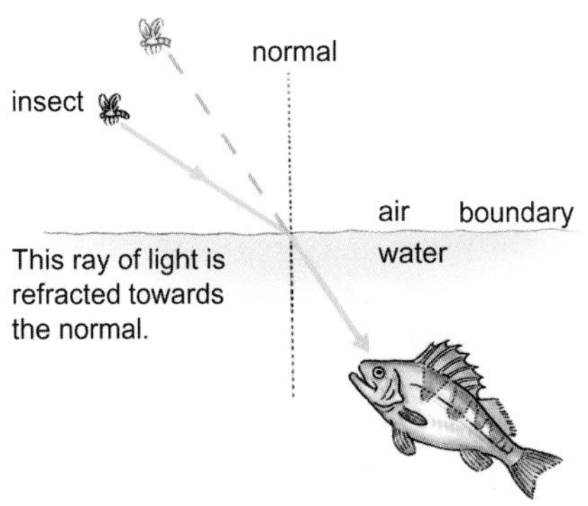

To the fish, the fly seems higher up than it actually is.

Question 5

8K.5 The spectrum

You should already know | Outcomes | Keywords

Newton's discovery

If you go to Woolsthorpe Manor in Lincolnshire, you can see a piece of glass called a **prism** set up to catch the Sun's rays coming through a hole in a window shutter.

The glass prism splits the sunlight into a rainbow of colours on the wall.

The room is the place where Isaac Newton first worked out that white light is made up of colours, on 21st August 1665.

The prism used to split white light into its colours has a triangular cross-section.

The prism refracts the light twice in the same direction. It splits white light up into the colours it is made from. The diagram shows you what happens.

ray box
beam of white light
prism made of clear plastic or glass
screen
spectrum of colours

Question 1

The spectrum

The rainbow of colours is called a **spectrum**. If you look at a spectrum, you will see that the colours gradually blend from one to the other, from a deep red at one end to a deep violet at the other.

One way to remember the order of the colours is to use this phrase:

Richard Of York Gave Battle In Vain.

Each capital letter stands for a colour.

Word	Richard	Of	York	Gave	Battle	In	Vain
Colour	red	orange	yellow	green	blue	indigo	violet

The seven-word phrase is just a way of remembering the order.

There are not actually any separate bands of colour in a spectrum, just a gradual change of shade.

Question 2

Rainbows

Sometimes you can see a rainbow when the Sun is shining and it is raining at the same time.

The raindrops work in a similar way to tiny prisms in the sunlight even though they are spherical and not triangular.

If you are going to see a rainbow, you have to be in the right position. The angles have to be correct between the Sun, the rain and your eyes. You need

- to stand with your back to the sunlight
- to make the angle between the direction of the sunlight and your line of sight to the raindrops about 42°.

When you see a rainbow, red appears at the top of a rainbow and violet appears at the bottom.

The white sunlight enters the top of the raindrops. It splits into a spectrum. Each colour comes out at a different angle. Red light comes to you from the higher raindrops. Violet comes to you from the lower ones. The spectrum is formed by the colours coming out of the raindrops in between.

You can make an artificial rainbow with a mist of drops from a garden hose.

Drops of rain can be split into all the colours of the rainbow.

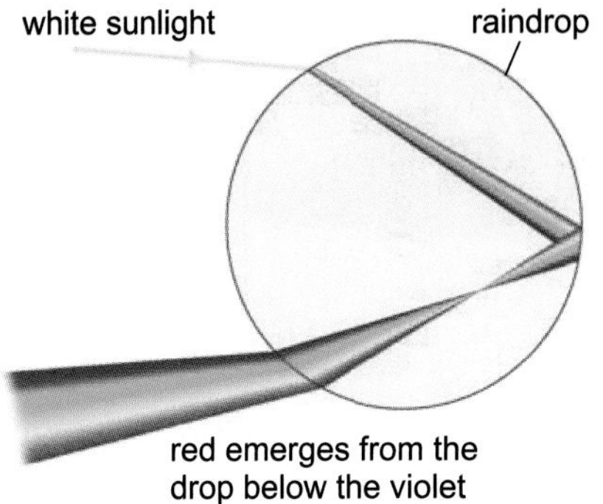

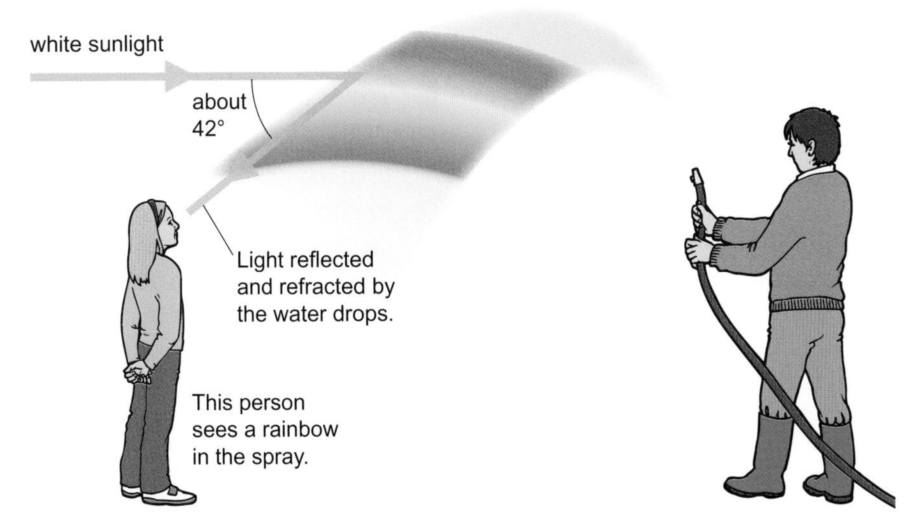

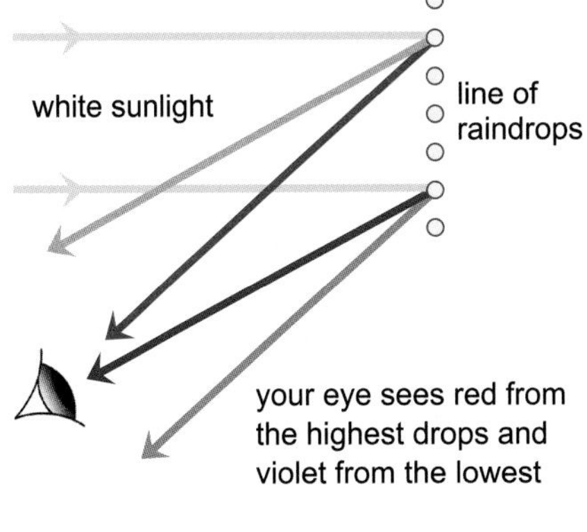

Another way of splitting white light into colours is to look at the Sun's reflection on the shiny side of a CD and to tilt the CD at the same time.

Question 3 | 4

Review your work

Summary ➡

8K.HSW Using light in different ways

You should already know | Outcomes | Keywords

Using light

We get most of our information about the world using light by looking at things.

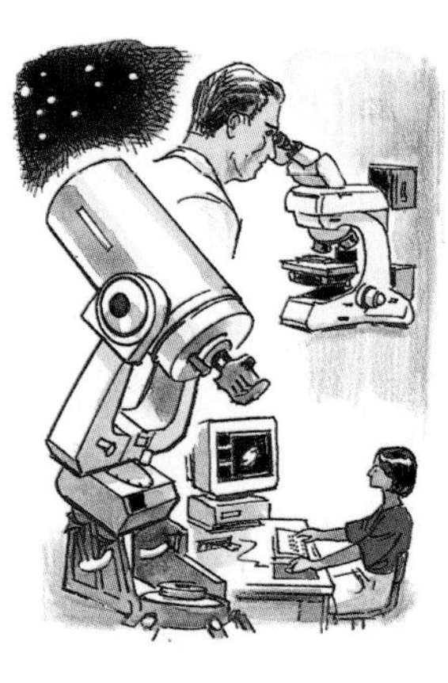

The diagrams show some of the ways we use light.

These are examples of how scientific ideas have been applied to bring about **technological developments**.

Some of them have happened within the past hundred years. The first traffic lights were installed in London in 1932.

Laser light

Laser light is very pure light. This light was first made in 1960.

Since then, a lot of development has happened. Many scientists have shared ideas. This is an example of scientists working together.

Laser light can be made to carry so much energy that it will cut steel.

It can also be made so that it carries less energy and is safe to use in everyday life.

- It is used to scan bar codes in supermarkets.
- It is used to read the information on CDs and DVDs.

Laser light is reflected from the white parts of the bar code and absorbed by the black bars. A sensor detects the reflected light pattern as the code is scanned. The pattern depends on the widths of the bars. A computer translates the pattern into a number that identifies the product.

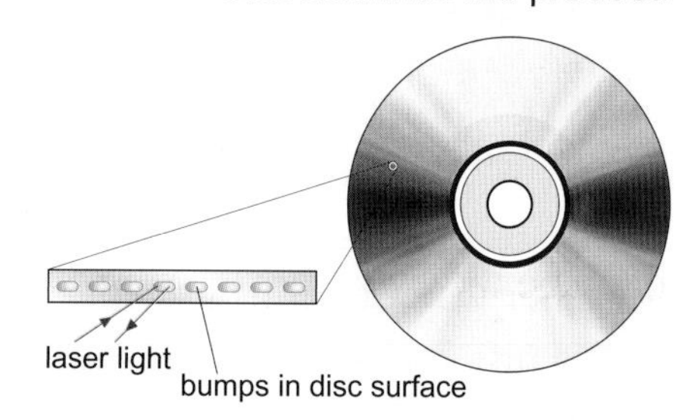

A CD player works in a similar way. A pattern of microscopic bumps in the disc reflect the laser beam as the disc spins.

Question 1  

Optical fibres

Optical fibres were invented in 1955. An optical fibre is a very fine fibre that is transparent. Light does not escape out of the sides of the fibre. This is a very useful idea.

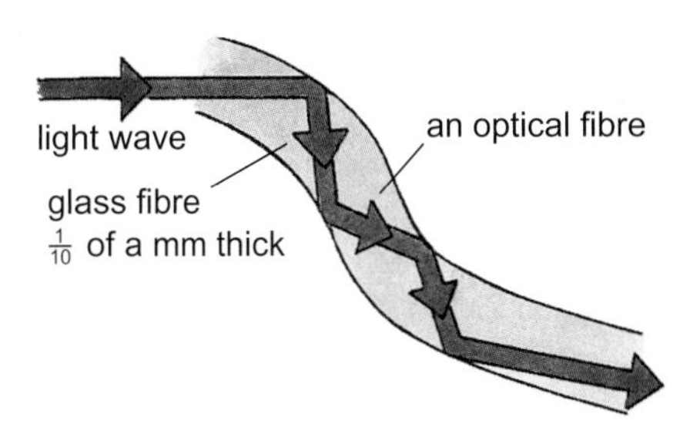

Light waves travelling through an optical fibre.

We use optical fibres in medicine.

Optical fibres let medical staff see what is happening inside patients without cutting them open. This is done using a group of optical fibres called an endoscope. The diagrams show how it works.

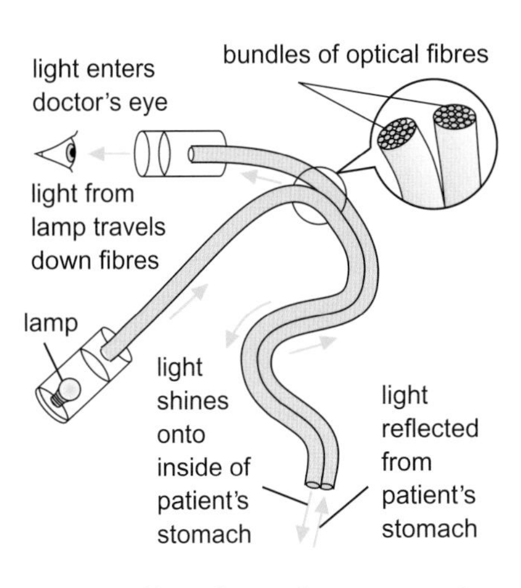

How the endoscope works.

Optical fibres are used by surgeons to carry out keyhole surgery. This is safer than major surgery. It causes less damage to the patient.

This is an example of applying a scientific idea to bring about changes in people's lives.

It is also an example of people sharing developments between different areas. In this case, scientists in physics and medicine are working together. This is called **collaboration**.

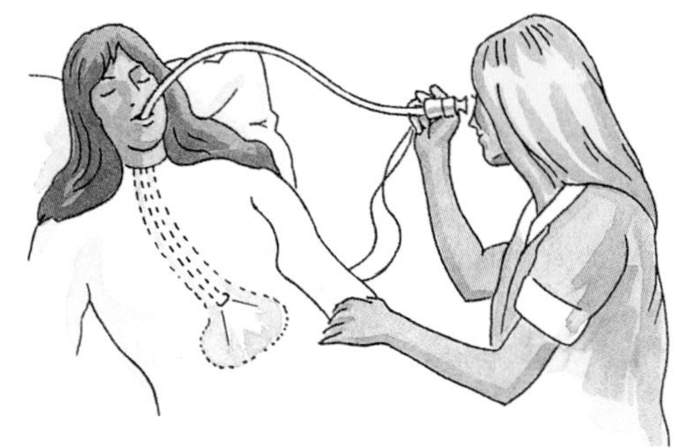
Using an endoscope ('endo' means 'inside' and 'scope' means 'looking').

Question 4 | 5

Safety

Scientists assess the risk of what they do in the laboratory and the workplace.

You can see **hazard** warnings like this on equipment like CD players.

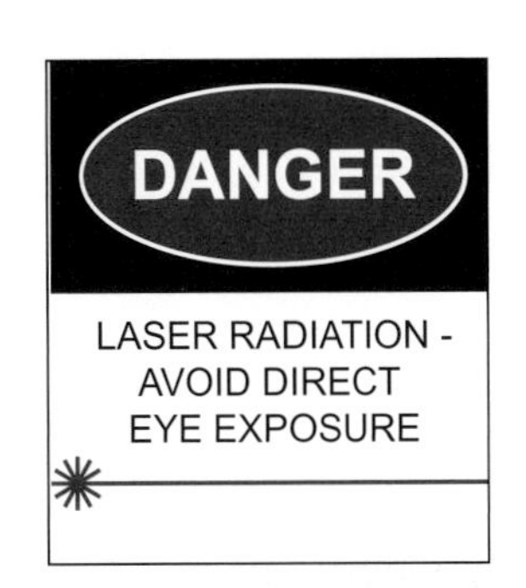

Laser light is not the only type of light that can be dangerous. Any very bright light can damage the eye.

A welding torch makes metal white hot. The operator must use a mask with a dark filter to protect his eyes from the high intensity light.

Question 6

8K Questions

8K.1

1. How can you show that light travels in straight lines?
2. Why does a pencil make a shadow when you hold it in front of a torch?
3. Why do you see lightning before you hear thunder?
4. Why did early attempts to measure the speed of light fail?

8K.2

1. What three things can happen to light when it hits a substance?
2. **a** What is meant by a transparent substance?
   b Why is a car windscreen made from a transparent substance and not a translucent one?
3. **a** What is meant by a translucent substance?
   b Why is a bathroom window made from translucent glass not transparent glass?
4. **a** What is meant by an opaque substance?
   b Why is a film canister made from an opaque substance?
5. What is the difference between what happens when light things and dark things reflect light?
6. What has to happen for you to see a non-luminous object?
   Give one reason for your answer.

8K.3

1. Explain what is meant by these terms when light reflects off a plane mirror.
   a Angle of incidence
   b Angle of reflection
   c Normal
2. What is the law connecting the angle of incidence and the angle of reflection when light reflects off a plane mirror?
3. When you look in a mirror, how far away does the image seem to be?
4. Give two examples of the use of plane mirrors in everyday life.

8K.4

1 What is meant by <u>refraction</u> and when does it happen?

2 When light goes into a substance like water or glass from the air, does it bend towards the normal or away from the normal?

3 When light comes out of a substance like water or glass into the air, does it bend towards the normal or away from the normal?

4 What is the only angle that light can travel along for there to be no bending?

5 Describe <u>three</u> everyday effects of refraction.

8K.5

1 What does a prism do to white light?

2 What are the <u>seven</u> colours of the spectrum?

3 What do you have to do to see a rainbow?

4 Give <u>two</u> other examples of where you can see a spectrum (apart from when it is raining and sunny at the same time, and by using a prism).

8K.HSW

1 Give <u>three</u> examples of how technological developments in the use of light have changed the way people behave.

2 In 2003, a town in Holland called Drachten stopped using its traffic lights, and the number of road accidents fell. In September 2007, the London Borough of Kensington and Chelsea announced that it had plans to do the same thing in some streets.

  a Suggest how traffic lights change motorists' behaviour. Why might removing traffic lights reduce the number of accidents?

  b Use the Internet to research the development of traffic lights from 1868 to the present day.

3 Find out what the letters of the word 'laser' stand for, and find out how laser light is used to read a CD.

4 Optical fibres are used in communication systems. Find out how an optical fibre can be use to replace a metal wire in a telephone system.

5 Laser light can be used to change the shape of a person's eye so that they do not need to wear glasses or contact lenses. This is sometimes called 'laser eye treatment'.
Use the Internet to find out what this is. Make a list of points for it and against it.

6 Find out which part of the eye is likely to be damaged if high-energy light like laser light enters your eye.

8L.1 Making different sounds

Scientific enquiry

You should already know

Outcomes

Keywords

What causes sound?

Things make sounds **vibrate**.

Sounds are caused by **vibrations**.

A vibration is a fast backwards-and-forwards movement that repeats many times. If you put your fingers against your throat as you speak, you will feel the vibrations.

A saxophone has a reed in the mouthpiece that vibrates when it is blown. The moving reed makes the air inside the tube of the instrument vibrate. It produces a note.

Musical instruments have some way of making vibrations. These make the sounds and notes that the musician needs.

A loudspeaker makes a sound when its paper cone vibrates backwards and forwards.

You can hear the washing machine because it is vibrating.

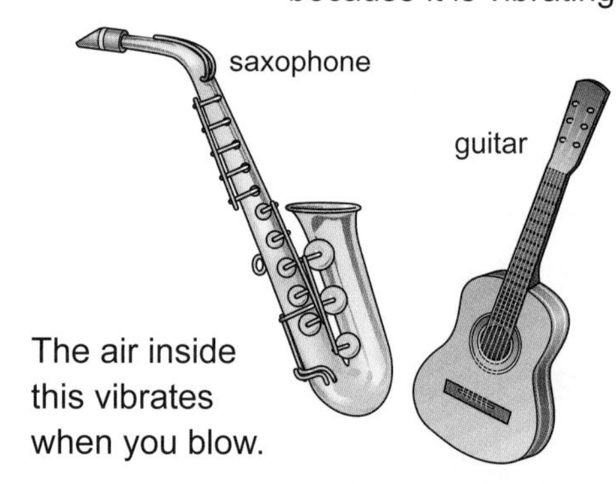

The air inside this vibrates when you blow.

The strings vibrate when you pluck them.

Question 1 | 2

Different sounds

Some sounds are loud. Some sounds are soft.

We refer to this as their **loudness**.

The loudness of a sound you hear depends on three things:

- how big the vibration is that produces the sound
- how far away you are from the sound
- if there is anything between you and the sound.

The photographs show two situations in which you make the sound louder by making the vibrations larger.

Hitting the drum harder produces a bigger vibration.

Turn up the volume to produce a bigger vibration in the loudspeaker.

Question 3

What is amplitude?

Loud sounds are made by vibrations that are large. We say that the vibrations have a large **amplitude**.

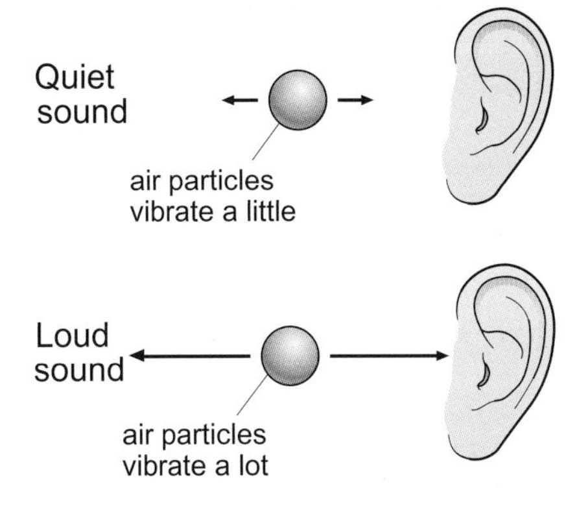

The amplitude of a vibration is how far something moves from its rest position when it moves back and forward.

- When the sound is loud, the air particles vibrate a lot. They have a large amplitude.
- When the sound is quiet, the air particles only vibrate a little bit. They have a small amplitude.

Question 4

What is pitch?

Sounds can be high or low. A motorbike engine ticking over gives out a sound with a low pitch. A cat complaining about something gives out a high-pitched sound.

The pitch depends on how many vibrations the source of the sound makes in a second. This is called the **frequency**. Frequency is measured in **hertz** (symbol Hz).

- A frequency of 1 Hz means that there is one vibration per second.
- 2000 Hz means that there are 2000 vibrations per second.

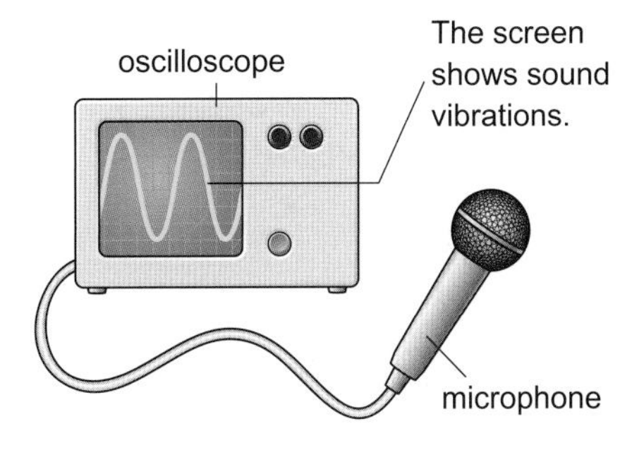

Big objects usually produce lower frequencies than smaller objects because they vibrate more slowly than small objects.

The picture shows a violin and a cello. If both instruments are played in the same way, the note from the violin will be higher.

Question 5 **6**

Looking at pitch and amplitude

If you connect a microphone to an oscilloscope, you can get a picture of a sound. The microphone changes the sounds into an electrical signal. The oscilloscope shows the electrical signal as a graph.

- The height of the oscilloscope signal shows the amplitude.
- How often the wave goes up and down gives the frequency.
- We measure amplitude from the mid-point to the highest point of a wave.

This is the amplitude of the wave.

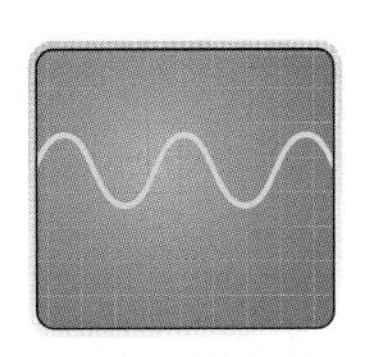

Quiet sound.

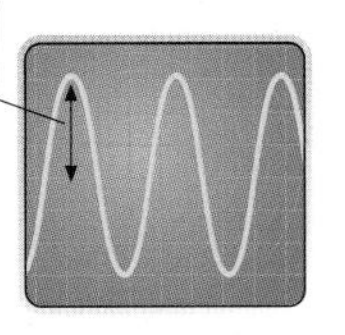

Loud sound.

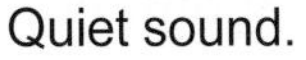

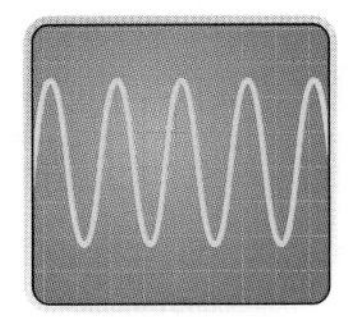

High-pitched sound.

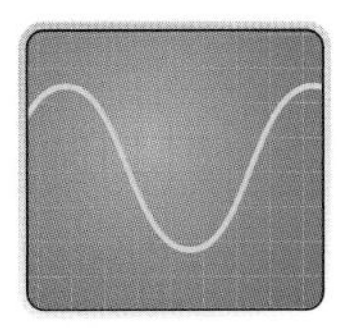

Low-pitched sound.

Question 7

8L.2 Detecting sound

You should already know | **Outcomes** | **Keywords**

How sound travels from a loudspeaker

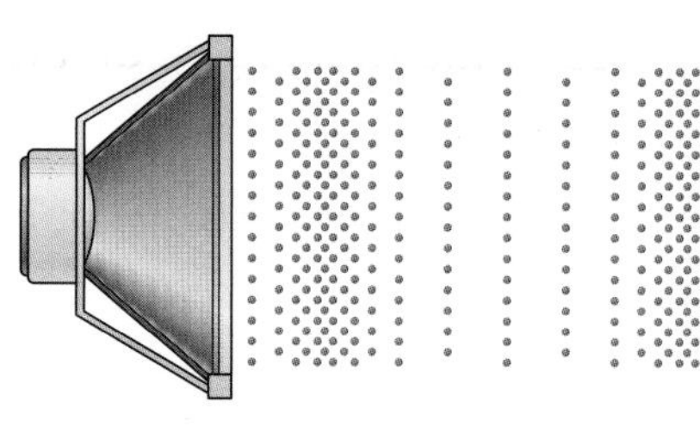

When a loudspeaker produces a sound, its cone vibrates backwards and forwards.

This makes the air particles in front of it vibrate.

- The vibration is passed from air particle to air particle.
- It is called a sound wave.

If you put a candle flame in front of a loudspeaker, the flame will vibrate backwards and forwards.

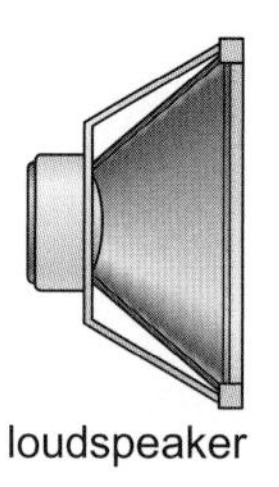

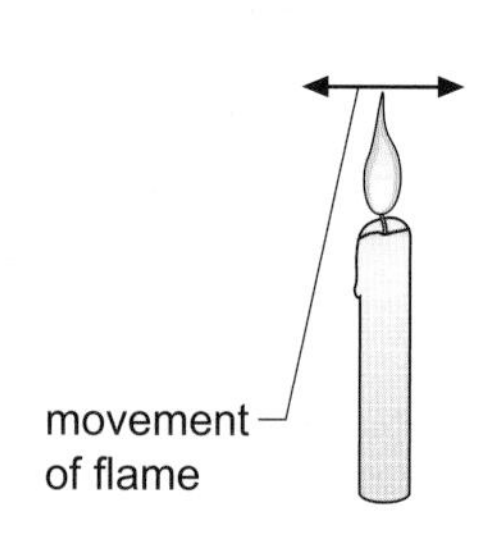

Question 1

Sound and a vacuum

Sound reaches you because the air particles transfer it in a wave. Sound cannot travel if there is nothing to pass the vibration along.

This effect was first shown in 1705 by a scientist called Francis Hauksbee. He used a clock in a jar with the air removed. When the air was removed, there was no sound from the clock.

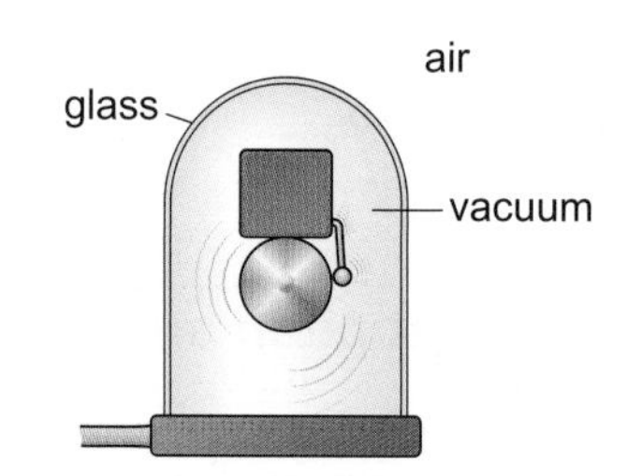

The diagram shows the same idea using a bell. When all the air has been sucked out of the glass jar, we say there is a **vacuum** in the glass jar. The bell cannot be heard if it is in a vacuum.

Sound cannot travel through a vacuum because there are no particles to make a sound wave.

Once you leave the Earth's atmosphere, you are in the vacuum of space. Astronauts use radio waves to communicate when they are in space. Radio waves are like light – they will travel through a vacuum.

No sound can be heard.

Question 2

The speed of sound

You can measure the speed of sound with a gun that produces a lot of smoke and a very loud sound. The cartoon shows you how. It is safer if the gun is loaded with a blank cartridge instead of live shot!

To work out the speed of the sound, you divide the distance the sound travels by the time it takes to travel.

Accurate experiments show that sound travels at 330 m/s in still air.

The speed of sound is affected by things like the wind speed and how much water vapour is in the air.

The speed of sound is different in different materials. Where particles are closer together, the sound vibrations can be passed along much more quickly.

Material	Speed of sound in m/s
air	330
water	1500
brick	3000
iron	5000

Some typical values for the speed of sound in various materials.

Question 3 4

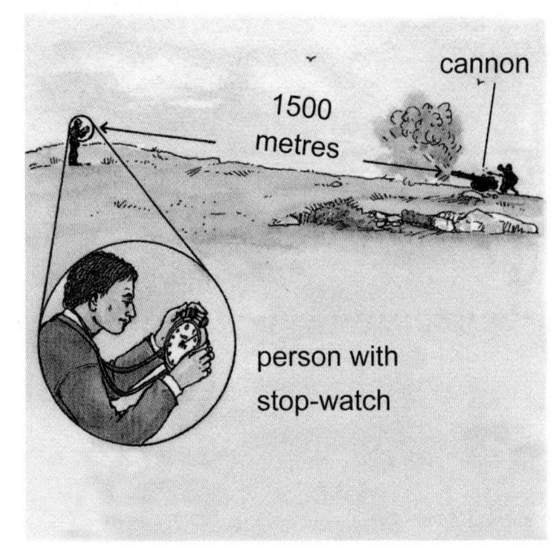

The person with the watch starts it when they see the smoke. They stop the watch when they hear the sound. The watch shows 5 seconds.

Reflections

Sound will reflect off hard surfaces in a similar way to light. You can show this with two tubes and a ticking watch. The diagram shows you how.

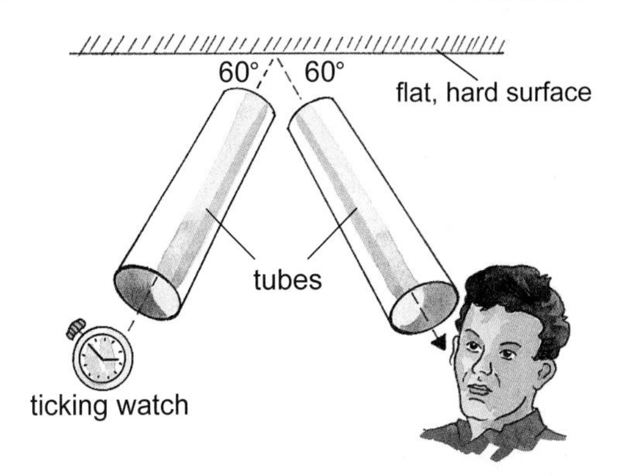

The ticks sound loudest when the tubes are at the same angle to the surface.

Because sound reflects off hard surfaces, things sound different inside concert halls than they do in the open air.

Concert halls produce sound reflections. These are called **reverberations**.

Reverberation is the name for the very faint echoes you hear from different parts of the hall, reaching you at different times.

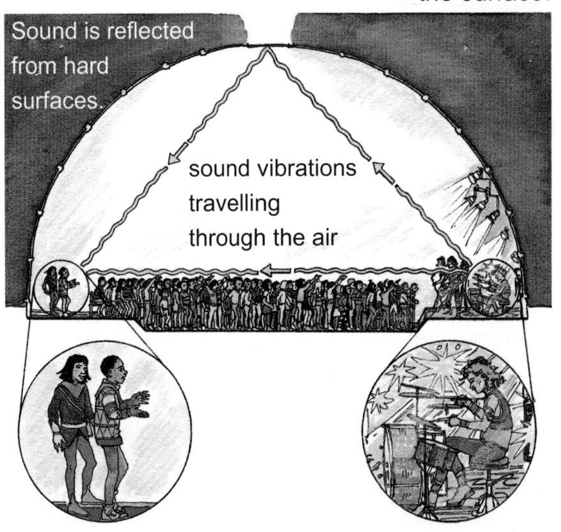

The sound from the drummer reaches the people at the back by different routes.

Question 5

Check your progress

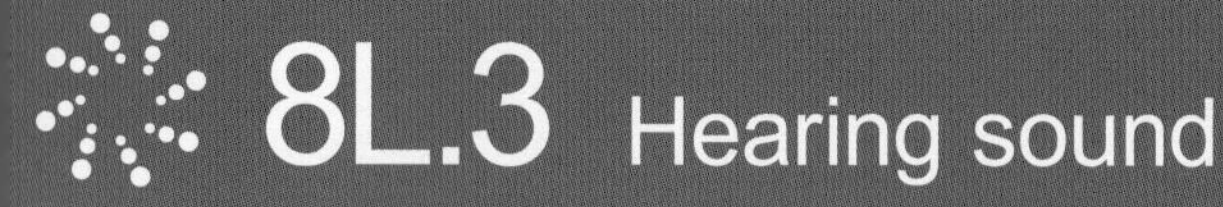

8L.3 Hearing sound

You should already know | Outcomes | Keywords

The ear

The **ear** is the part of the body that converts the vibrations of sound to the messages in nerves that go to the brain.

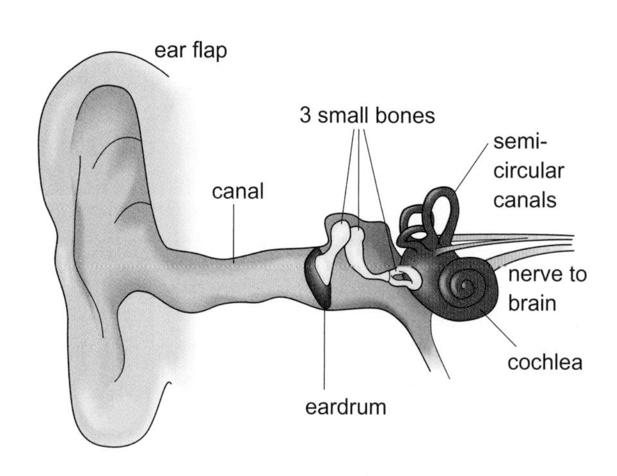

- A sound wave travels down the ear canal and makes the **eardrum** vibrate.
- These vibrations are passed on to the **cochlea** by a set of three small bones.
- In the cochlea, a liquid moves backwards and forwards, and stimulates the nerve cells inside it.
- The nerve cells make small electrical signals.
- These electrical signals travel along the nerve to the brain.

Question 1

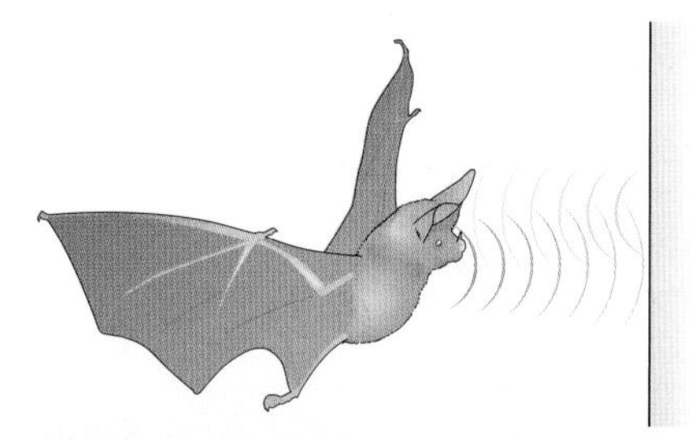

Animals have two ears. Having two ears helps them to work out the direction that a sound is coming from. This can be useful when catching prey and also when escaping from danger.

When animals generate sounds, the sound waves reflect off objects near them. Some animals detect the reflected sound and use the reflection to find their way about.

This is called **echo-location**. It is how a bat can fly about in the dark and not bump into things. The bat also uses it to locate its prey.

Long-eared bat.

Bats have large ears to collect faint sounds.

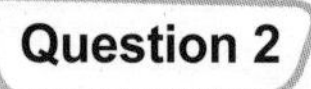

Question 2

Range of hearing

Different animals hear different ranges of sound. The highest note that a human aged about 20 years old can hear has a frequency of about 20 000 Hz.

Dogs, cats and bats can hear much higher notes than humans.

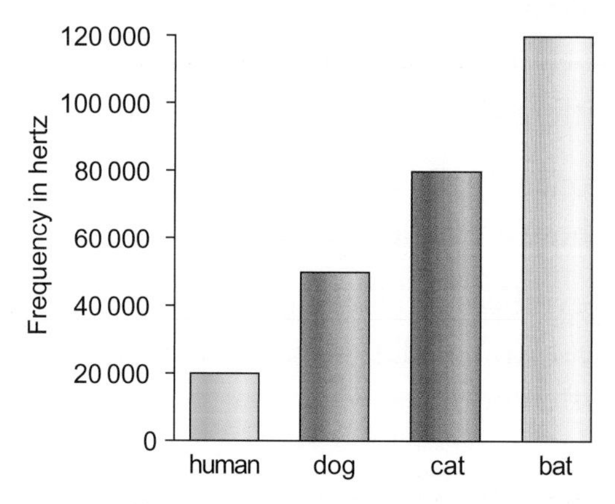

Hearing ranges of different animals.

Dog owners sometimes use a special whistle. This produces a note that is too high for humans to hear, but the dog can hear it.

We say that a sound that is higher than the limit of human hearing is ultrasonic. The dog whistle in the photograph produces an **ultrasonic** sound.

Question 3 | 4

Damage to hearing

Loud sound waves carry a lot of energy. Very loud sounds can hurt and even do permanent damage.

This can happen

- at a loud rock concert
- if you play music too loudly through headphones.

The loudness of a sound is measured in a unit called the **decibel**. It has the symbol dB.

The diagram shows some typical sounds. Those higher than 90 dB can cause permanent damage to your hearing if you keep being exposed to them.

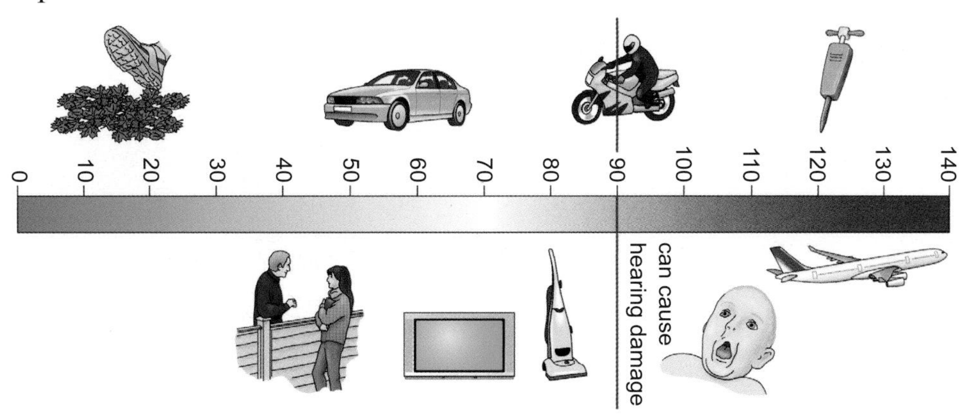

Too much noise is called **noise pollution**. People living near airports suffer from noise pollution when aircraft take off and land.

There are strict laws to control noise pollution.

There are also regulations to protect workers who do noisy jobs.

The worker with the road drill in the photograph is wearing ear protectors, so the sound that he hears from the drill is not over 90 dB.

Question 5 | 6

Review your work

Summary ➡

8L.HSW Using sound

You should already know | Outcomes | Keywords

How good is the experiment?

In science, you need to decide if the method is good enough to give you the evidence you want.

We say you need to **evaluate** the method. In the experiment below, some things could be improved.

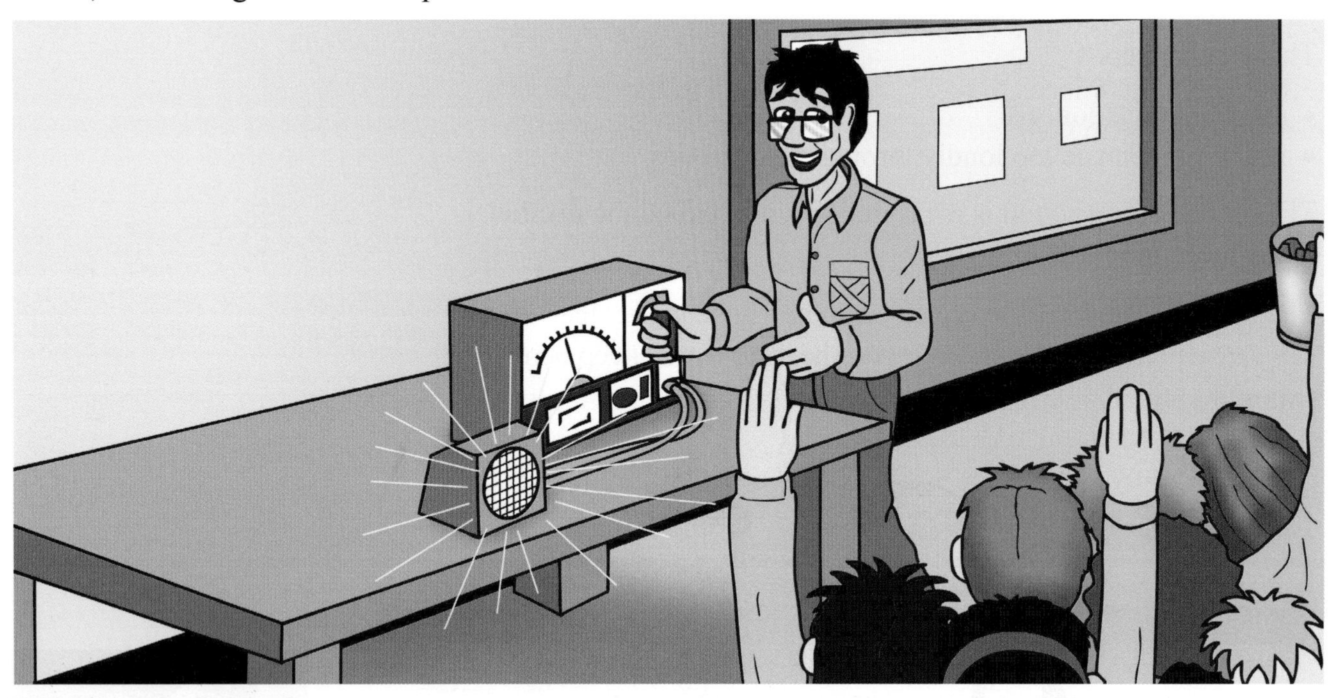

The aim is to find out the highest frequency that pupils can hear.

- The teacher increases the frequency of the sound produced.
- The pupils put their hands up when it becomes too high to hear.

The teacher asks these questions to help pupils evaluate the experiment.

- "Does it matter that you are sitting at different distances from the loudspeaker?"
- "Might you be influenced when you see others put their hands up?"
- "Not all of you put your hand up at the same time. What is the best way of recording the results to take this into account?"

Question 1 

Applying science

Using scientific ideas can help to improve people's lives.

The high-pitched sounds that are above the human hearing range are called ultrasound. Ultrasound is used to improve people's lives.

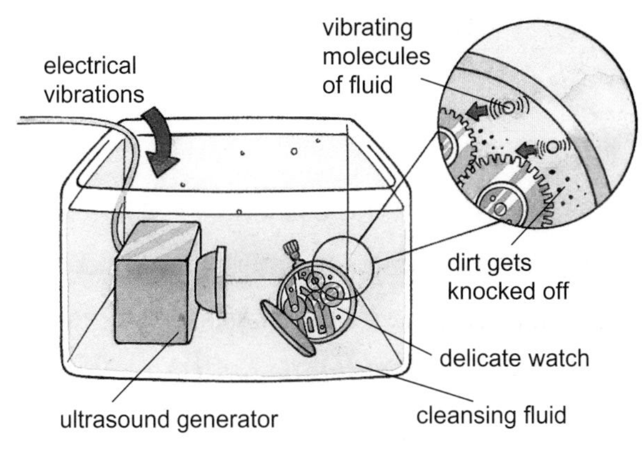

Cleaning

A generator can be used to make ultrasound vibrations in a bath of liquid. You can use this to clean a delicate thing like a watch. The vibrations of the sound make the molecules of the liquid knock the dirt off.

Dentists use a similar system for cleaning some of the small tools they need to use in your mouth.

Scanning

Ultrasound is used to scan someone who is pregnant to look at the fetus. This lets a doctor see how the woman and fetus are progressing.

- A probe is moved over the woman's body.
- The probe gives out ultrasound.
- The probe also detects ultrasound that is reflected from the fetus inside the woman's uterus.
- The detected ultrasound is fed into a computer and the software produces a picture on the screen.

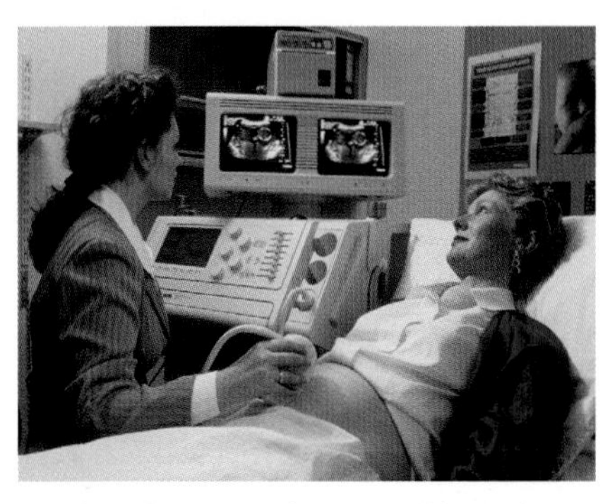

An ultrasound scan taking place.

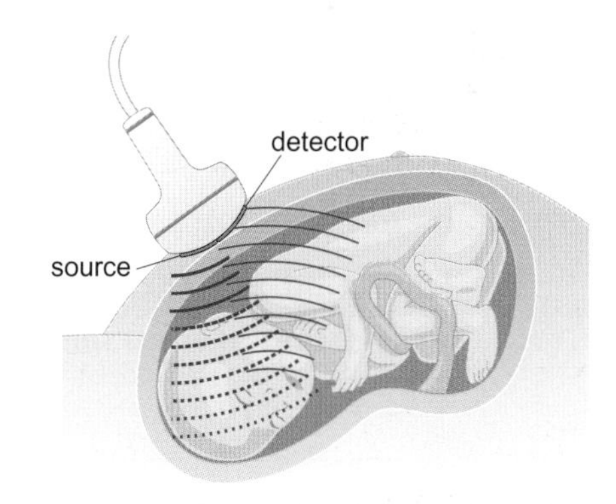

This is a good example of the result of **collaboration** between scientists who work in the field of sound and computer software engineers.

For most people, this is the first picture of the baby for the family album. It can also give a family an early warning that they are expecting twins, triplets or even more!

Unfortunately, it can have **moral and ethical issues** associated with it.

It is possible to tell from a scan what gender the fetus is. In some parts of the world, if the fetus is not the right gender, parents will consider killing it by having it aborted.

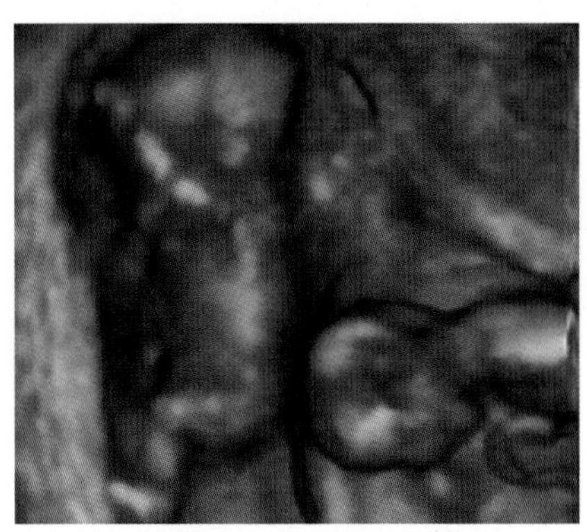

Question 3 4

8L Questions

8L.1

1. What is sound caused by?
2. What vibrates to cause the sound on a guitar?
3. What can you do to make a sound louder?
4. How does amplitude affect the loudness of a sound?
5. What is the connection between pitch and frequency?
6. The note middle C on the piano has a frequency of 256 Hz.
   What does this mean about how often the piano string vibrates?
7. How does an oscilloscope connected to a microphone show how loud a sound is?

8L.2

1. A candle flame will vibrate backwards and forwards in front of a loudspeaker.
   What does this show you?
2. **a** What is the name for a space from which almost all the particles of gas have been removed so that it is virtually empty?
   b Why doesn't sound travel through it?
3. How could you measure the speed of sound?
4. **a** In which type of substance (solid, liquid or gas) does sound travel fastest?
   b Suggest one reason why this is so.
5. The sound of someone singing or playing an instrument in the open air is different from the sound you get in a concert hall.
   Why is this?

8L.3

1. Describe how your ear detects sound.
2. Why is it useful for animals to have sensitive hearing?
3. Name two animals that can hear sounds that are above the limit of human hearing.
4. What does the term ultrasonic mean?
5. a What is noise pollution?

   b Suggest things that could be done to reduce it.
6. a How can loud sounds be dangerous?

   b What can be done to reduce the danger?

8L.HSW

1. Imagine that you are in the class where the experiment is taking place.
   Write out a set of instructions for someone to follow so that they can carry out the experiment. Include a way of displaying the results.
2. Look at the three questions asked by the teacher.

   a Suggest answers to the questions.

   b How good do you think the experiment was?

   c Suggest ways of improving it.
3. Describe two scientific developments in sound that have resulted in improvements in the ways that people live their lives.
4. What moral and ethical issues are associated with ultrasound scans of people who are pregnant?

8A Summary

Keywords

To stay healthy we need a balanced diet.

The amount of each food group that we need depends on:

- our age
- our size
- our sex
- how active we are.

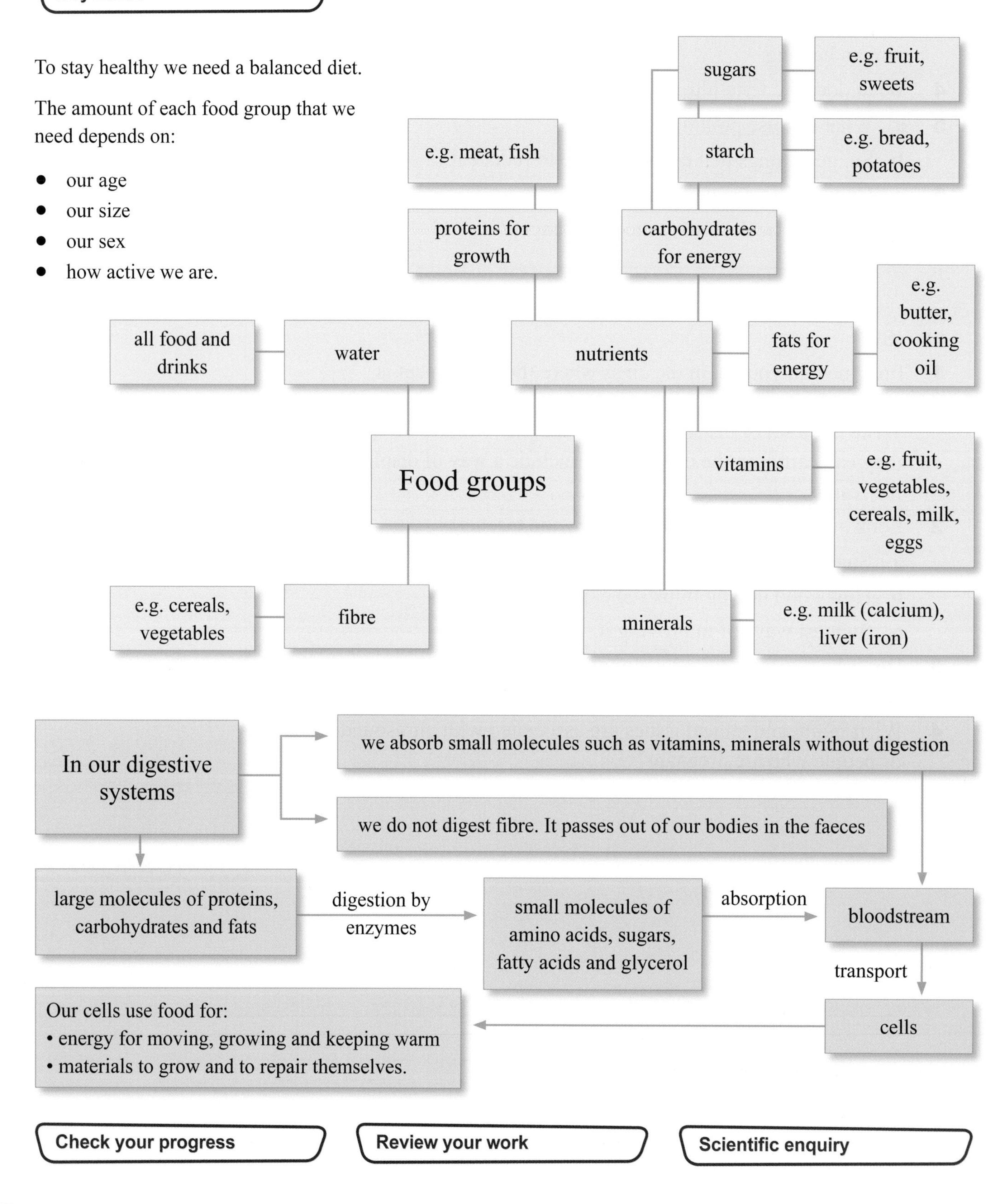

Check your progress **Review your work** **Scientific enquiry**

8B Summary

Keywords

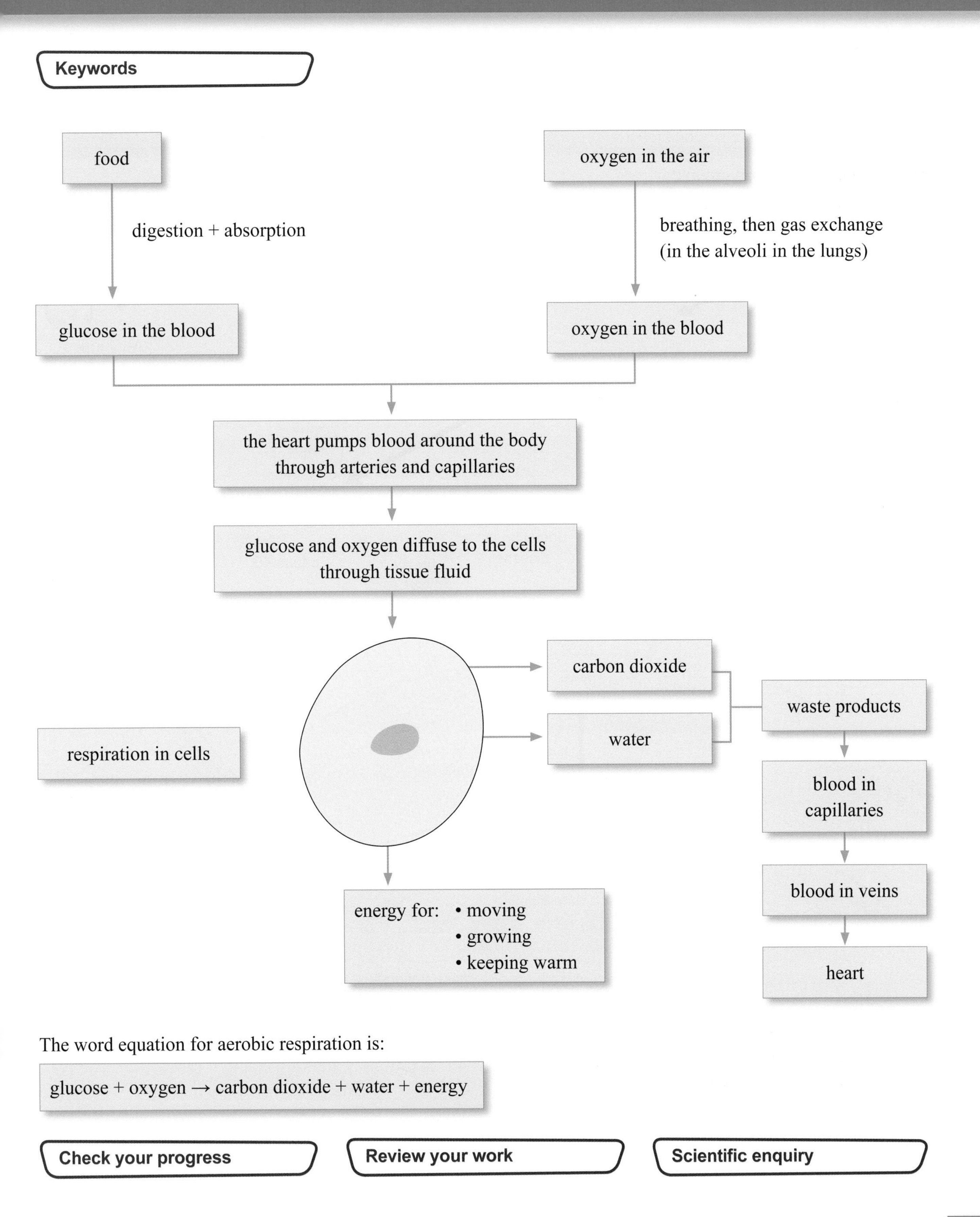

The word equation for aerobic respiration is:

glucose + oxygen → carbon dioxide + water + energy

Check your progress

Review your work

Scientific enquiry

8C Summary

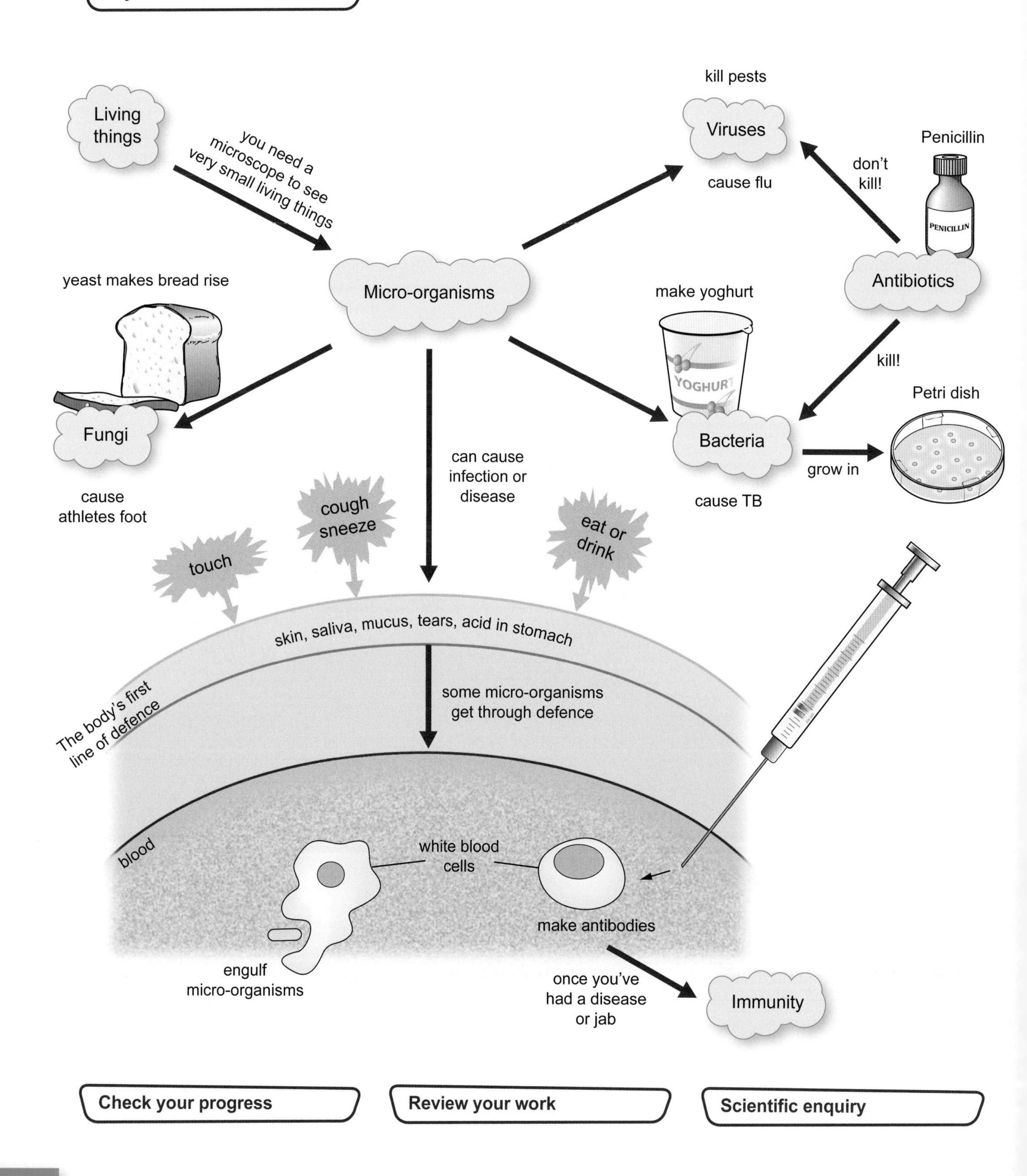
Keywords
Living things
you need a microscope to see very small living things
Micro-organisms
kill pests
Viruses
cause flu
Penicillin
don't kill!
PENICILLIN
Antibiotics
kill!
yeast makes bread rise
Fungi
cause athletes foot
make yoghurt
YOGHURT
Bacteria
cause TB
grow in
Petri dish
can cause infection or disease
touch
cough sneeze
eat or drink
skin, saliva, mucus, tears, acid in stomach
The body's first line of defence
some micro-organisms get through defence
blood
white blood cells
engulf micro-organisms
make antibodies
once you've had a disease or jab
Immunity
Check your progress
Review your work
Scientific enquiry

8D Summary

Keywords

Animals with backbones are called vertebrates.

Animals without backbones are called invertebrates.

We classify plants and animals into smaller groups.

Plants are divided into two groups: those with a transport system (vascular) and those without a transport system (non-vascular).

Some factors that affect behaviour are internal (e.g. hormones). Others are external (e.g. tides).

The behaviour of animals affects their survival.

Some behaviour is innate, some is learned.

The place where a plant or animal lives is called its habitat.

Plants and animals have features that suit them to where they live. We say they are adapted to their environmental conditions.

The plants and animals in a habitat interact.

A collection of species living in an area is called a community.

A group of organisms of the same species in an area is called the population.

To find the population size of an organism in an area we can use a quadrat.

An organism that makes its own food is called a producer.

Living things in a community depend on each other.

An animal that only eats plants is called a herbivore.

An animal that cannot make its own food, but gets it from other plants or animals is called a consumer.

An animal that feeds on other animals is called a carnivore.

A number of food chains joined together is called a food web.

Food chains show what animals eat.

A pyramid of numbers is a pyramid shaped diagram showing the numbers of organisms at each stage of a food chain.

Check your progress

Review your work

Scientific enquiry

8E Summary

Keywords

Key ideas

- An element is a substance that is made from just one type of particle, called an atom.
- There are about 100 elements.
- Most of the elements are metals.
- Each element has its own symbol.
- Elements combine to make different materials.
- Materials are used to make objects.
- Some elements were known in ancient times, but other elements have been discovered more recently.
- Information about elements is collected together in the periodic table.
- There are vertical columns in the periodic table called groups.
- Elements in the same group have similar properties.
- Atoms can combine to make molecules.
- Two or more different elements combine to form a compound.
- A chemical change makes new substances.
- New materials made in a chemical change have different properties from the substances they are made from.
- Chemical changes are different from physical changes.
- A physical change does not make new substances.

Check your progress

Review your work

Scientific enquiry

8F Summary

Keywords

Key ideas

- A compound contains different types of atom joined together.
- The formula of a substance tells us the proportion of each type of atom present in the substance.
- A compound has a chemical name like 'sodium chloride'.
- A chemical formula is used to describe the number of different atoms in one particle of a compound.
- There are different types of reaction, including combustion, neutralisation, precipitation and thermal decomposition.
- A mixture is formed when two or more substances are added together but do not react.
- Air is a mixture of gases that can be separated into pure substances.
- Air contains nitrogen, oxygen, argon, carbon dioxide and water vapour.
- Each of the gases found in air has important uses.
- Air can be liquefied and then the individual gases can be separated by fractional distillation.
- Sea water and mineral water are other examples of mixtures.
- Elements and compounds melt and boil at certain temperatures.
- Mixtures do not melt or boil at one particular temperature.

Check your progress

Review your work

Scientific enquiry

Keywords

Key ideas

- Rocks are made of a mixture of mineral grains.
- Non-porous rocks do not let liquids or gases pass through them.
- Porous rocks let liquids and gases pass through them.
- Sedimentary rocks are formed when sediments settle on the bottom of the sea.
- Limestone is an example of a sedimentary rock.
- Weathering breaks down rocks.
- Acid rain causes chemical weathering.
- Physical weathering is caused by changes in temperature.
- When water freezes in cracks, it expands and breaks off bits of rock.
- When the surface of a rock expands and contracts as temperature changes, it can crack.
- Plants and animals can cause weathering – this is called biotic weathering.
- When rocks rub against each other and wear each other away it is called abrasion.
- Weathering can produce rock fragments.
- Rock fragments are moved by gravity, wind and water.
- Rivers carry some rock fragments to the sea.
- Rock fragments are deposited as sediment.
- Hard parts of dead plants and animals can form fossils in sedimentary rock.
- The processes of weathering and the formation of sedimentary rocks happen very slowly.
- We measure the very long times needed for rocks to form on a scale called geological time.
- Scientists who study rocks are called geologists.

Check your progress

Review your work

Scientific enquiry

8H Summary

Keywords

Key ideas

- Sedimentary rocks can be changed by pressure and high temperature into metamorphic rocks.
- Sedimentary rocks are changed over millions of years as they are buried deeper inside the Earth.
- There is molten rock inside the Earth called magma.
- When magma cools, it solidifies into rock called igneous rock.
- Igneous rocks with large crystals are formed when magma cools slowly.
- Igneous rocks with tiny crystals are formed when magma cools quickly.
- When magma comes out of the Earth's surface, it is called lava.
- When lava comes out on the Earth's surface, it is called an eruption.
- A volcano can be formed from an eruption of lava.
- All types of rock can be broken down by the weather and eventually end up as material in sedimentary rock.
- The constant recycling of material through sedimentary, metamorphic and igneous rocks is called the rock cycle.

Check your progress

Review your work

Scientific enquiry

8I Summary

Keywords

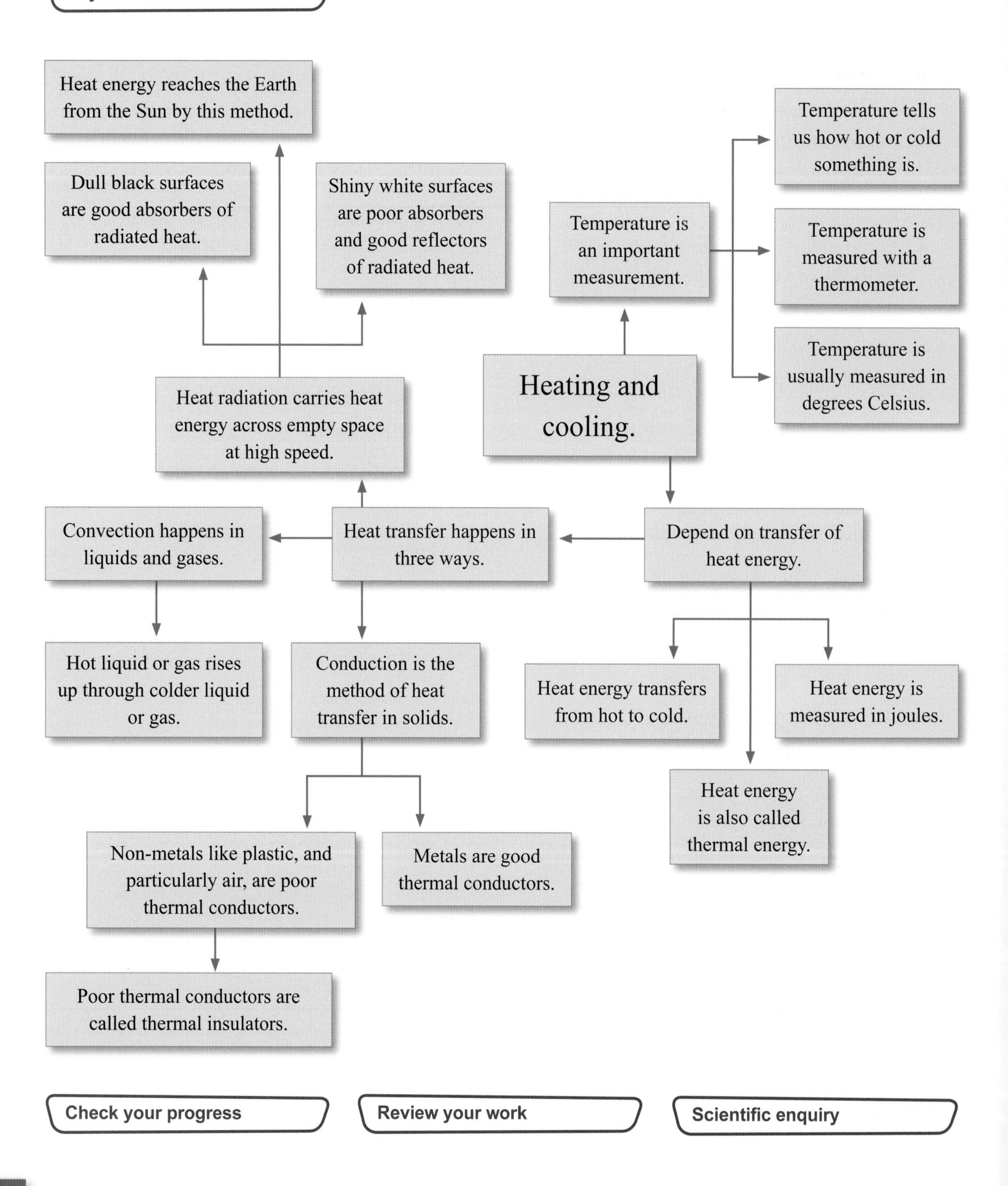

Check your progress

Review your work

Scientific enquiry

8J Summary

Keywords

Key ideas

- Magnets can attract and repel one another.
- Magnets attract magnetic materials.
- Examples of magnetic materials are iron, cobalt and nickel.
- Non-magnetic materials are not affected by magnets.
- The magnetic force passes through non-magnetic materials.
- The ends of magnets are called poles.
- The south-seeking pole points south and the north-seeking pole points north.
- A north-seeking pole and a south-seeking pole attract each other, but two south-seeking poles or two north-seeking poles repel each other.
- The area around magnets is called a magnetic field.
- Magnetic field lines can be plotted with a compass.
- The direction of the magnetic field goes from the magnet's north pole to its south pole.
- Stronger magnets are represented by showing more lines of force around them.
- The Earth behaves as if it had a giant magnet inside it.
- A compass contains a small magnet that is free to turn.
- The magnet in a compass lines up with the Earth's magnetic field.
- You can make your own magnet by stroking a piece of iron with one end of a magnet.
- When an electric current flows through a wire, there is a magnetic field around the wire.
- Magnets made using electricity are called electromagnets.
- An electromagnet consists of a coil of wire that carries an electric current.
- Electromagnets can be switched on and off.
- You can change the strength of an electromagnet by changing the size of the current, having more turns on the coil or placing a piece of iron called a core inside the coil of wire.
- Relay switches use the current from one circuit to switch on the current in another circuit.
- A relay contains an electromagnet.

Check your progress

Review your work

Scientific enquiry

8K Summary

Keywords

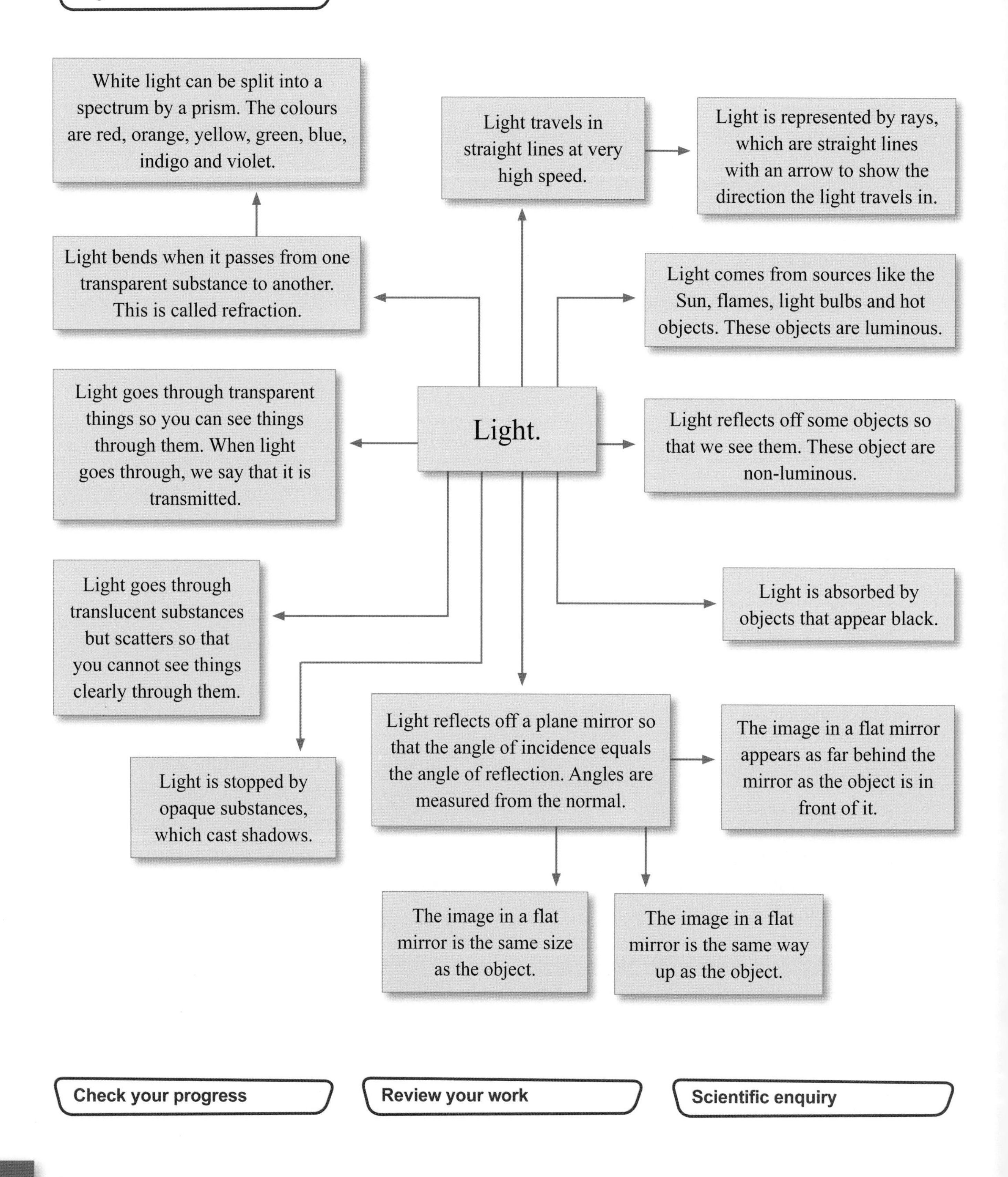

Check your progress | **Review your work** | **Scientific enquiry**

8L Summary

Keywords

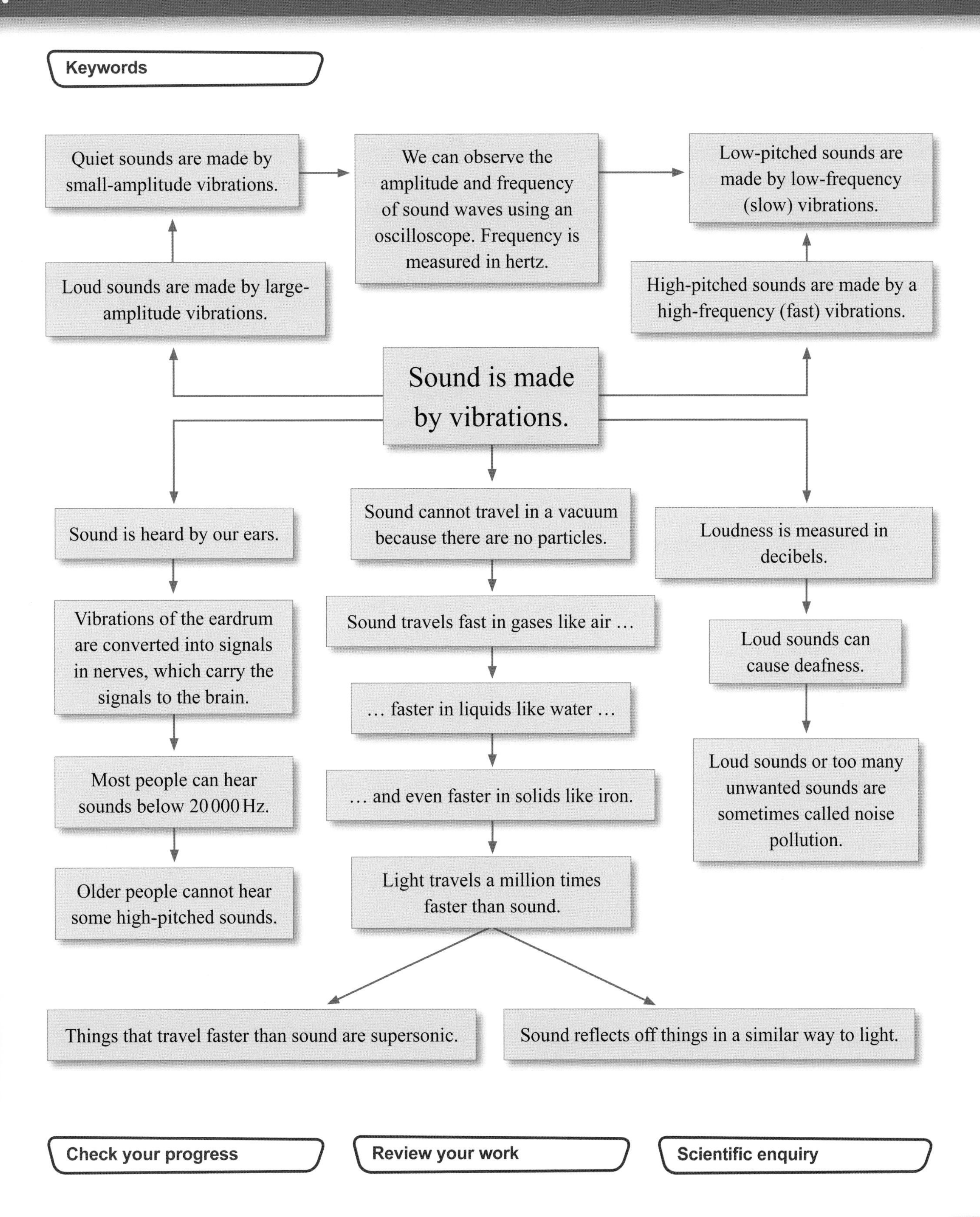

Check your progress

Review your work

Scientific enquiry

Glossary/Index

A

abrasion wear caused by one substance rubbing against another 100

absorb, absorption take in a substance; for example living cells take in oxygen and nutrients 9, 10, 12

absorbed (of light) when light falls onto something and is 'soaked up' by it so that it does not reflect or go through; an object that absorbs all the light that falls onto it will appear dull black 144

aerobic using oxygen 18, 22

alveoli tiny round parts that make up the air sacs in the lungs; one is an alveolus 24

amplitude the distance from the centre position of a wave to the peak; the loudness of sound 156

angle of incidence the angle between an incident ray and the normal 146

angle of reflection the angle between a reflected ray and the normal 146

antibiotics drugs used to kill bacteria in the body 42

antibiotic resistance able to grow in the presence of an antibiotic 44

antibodies chemicals made by white blood cells to destroy bacteria and other micro-organisms 42

arteries blood vessels that carry blood away from the heart 20

atmosphere the mixture of gases that surrounds a planet; on the Earth, the atmosphere is often referred to as 'the air' 86

atoms the smallest particle of an element 68

attract two objects pulling towards each other 130

B

bacteria micro-organisms that are cells without true nuclei; one is called a bacterium 36, 42

bacteriophages viruses that infect bacteria 44

behaviour patterns behaviour charateristics of members of a species 52

biased in the case of an opinion, based on someone's feelings rather than all the facts 60

biotic weathering the breakdown of rocks caused by plants and animals 98

breathe, breathing taking air in and out of the lungs 18, 24

breeding behaviour behaviour associated with breeding, such as courtship, mating or nest building 52

C

capillaries narrow blood vessels with walls only one cell thick 20

carbohydrates carbon compounds used by living things as an energy source, for example starch and sugars 2, 4, 8

carbon dioxide a gas in the air produced by living things in respiration, in combustion or burning and when an acid reacts with a carbonate 18, 28, 30, 86

carnivore an animal that eats flesh 58

cause a factor that causes a particular outcome is its cause; there has to be evidence for the factor being the cause not just a linked factor 26

Celsius a temperature scale with the melting point of ice written as 0 °C and the boiling point of water as 100 °C 118

chemical change a reaction between chemicals; it produces a new substance 70

chemical name the name used for a substance in chemistry; e.g. sodium chloride is the chemical name for table salt 78

chemical weathering when chemical reactions cause the weathering or breakdown of rocks 98

classify sort things into groups 50

cobalt a light-grey metal element that is attracted to permanent magnets 130

cochlea part of the inner ear; converts movement into electrical pulses which travel along nerves to the brain 160

collaborate, collaboration when people work together to carry out a task or solve a problem 26, 74, 152, 162

combustion when substances react with oxygen and release energy; another word for burning 80

community all the plants and animals that live in a particular place 54

compass a device which has a magnetic needle which points towards the North Pole 130

compound a substance made from the atoms of two or more different elements joined together 68, 70

conclusion the final result; the answer reached after looking at the evidence 104

conduction in heat conduction, energy passes along a solid as its particles heat up and vibrate faster 120

conflict where things oppose each other or point in opposite directions 74

consistent things are consistent when they fit in with each other and point to the same conclusion 104

consumer an animal that cannot make its own food, but eats plants and other animals 58

continental drift the idea that the continents are moving very slowly over the surface of the Earth 114

control part of an experiment that is needed to make a test fair; it is needed so that we can be sure of the cause of a change or a difference 30

convection a method of heat transfer in fluids (liquids and gases); the fluids at a higher temperature rise towards the top of the container they are in 122

convection current the flow of fluid when convection happens 122

critical making judgements concerning things that might be wrong, mistaken or not work 74

crystals a substance that forms from a melted or a dissolved solid in a definite shape 110

D

datalogger a piece of equipment that stores measurements for processing by a computer 126

decibel unit used to measure the loudness of sound 160

degrees intervals on a temperature scale 118

deposits, deposition when eroded rock fragments settle 100

diffusion the spreading out of a gas or a dissolved substance because its particles are moving at random 18

digestion the breakdown of large, insoluble molecules into small soluble ones which can be absorbed 9

disease when some part of a plant or animal isn't working properly 36, 42

E

ear the organ of hearing 160

eardrum a tightly stretch skin found at the end of the ear canal; it vibrates when sound reaches it 160

echo-location a method used to judge the distance to objects, used by bats. This involves sending out a high pitched sound which reflects off objects 160

egest get rid of faeces from the digestive system 12

electromagnet a device which becomes magnetic when an electric current flows through a wire coil 134

element a substance that can't be split into anything simpler by chemical reactions 66

emitter an object or surface that gives something out for example a hot surface will emit a lot of infra red radiation 124

energy energy is needed to make things happen 18, 30

environmental conditions conditions such as light level and temperature in the environment 54

enzymes protein substances made in cells; they speed up chemical reactions 10

epidemic an outbreak of a disease affecting a large number of people 40

epidemiology the study of the causes of disease 26

erupts when lava, volcanic ash and gases come out onto the surface of the Earth 110

ethical an ethical idea is one that relates to a set of moral principles 90

evaluate to consider how good an experiment or procedure was and how it could be improved 162

evidence observations and measurements on which theories are based 20, 104

expansion when a substance gets bigger because its particles speed up and move further apart 98

F

faeces undigested waste that passes out through the anus 12

fats part of our food that we use for energy 2, 4, 8

fibre undigestible cellulose in our food; it prevents constipation 4, 8, 12

food chain a diagram showing what animals eat 58

food web a diagram showing what eats what in a habitat 58

formula, formulae uses symbols to show how many atoms of elements are joined together to form a molecule of an element or a compound 78

fossils remains of plants and animals from long ago 102

fractional distillation the separation of a mixture of liquids by distillation 86

frequency the number of waves (for example sound waves) produced every second 156

fungi a group of living things, including micro-organisms such as moulds and yeasts which cannot make their own food; one is called a fungus 36, 42

G

gas exchange taking useful gases into a body or cell and getting rid of waste gases 24

geological time divisions of time based on the ages of rocks 102

geologists scientists who study the origin, composition and structure of rocks 96

global warming the idea that the temperature of the Earth is increasing 86, 90

glucose a carbohydrate that is a small, soluble molecule (a sugar) 18

greenhouse gases gases that contribute to the greenhouse effect; e.g. carbon dioxide and methane 86

groups vertical columns of elements in the periodic table; elements in the same group have similar properties 72

grow become bigger and more complicated 2

H

habitat the place where a plant or animal lives 50

hazard a source of danger 152

heart an organ that pumps blood 20

heat energy energy possessed by hot objects 120

herbivore an animal that eats plants 58

hertz unit of frequency 156

high temperatures a general term used to describe temperatures that are usually well above the boiling point of water 108

I

igneous formed when ash or molten magma from inside the Earth cools and solidifies 110

image the reflection of an object in a mirror 146

immunisation an injection given to give the patient immunity 42

immunity ability to resist an infectious disease as a result of immunisation or having had the disease 42

imprinted a word for young fixed on their mother, as when the young of some groups of birds instinctively recognise and follow their mother 52

infection disease caused by micro-organisms 39, 40

infra-red radiation waves that transfer thermal energy from an object; the higher the temperature of an object, more thermal energy is transferred 122

innate inborn or instinctive, so innate behaviour is behaviour that animals are born with 52

instinctive inborn or innate, so instinctive behaviour is behaviour that animals are born with 52

invertebrates an animal without a backbone 50

iron a silvery metallic element that is attracted to magnets 130

iron core a piece of iron used in the centre of an electromagnet 134

J

joules energy or work is measured in units called joules 120

L

large intestine the wide part of the intestine between the small intestine and anus 10

lava igneous rock that solidifies on the Earth's surface 110

learned behaviour behaviour that animals learn by trial and error or as a result of teaching 52

light source something which gives out light 142

loudness human perception of the intensity of a sound 156

luminous gives out its own light 144

lung organ for gas exchange between the blood and the air 22, 24

M

magma molten rock from below the Earth's crust 108, 110

magnetic field the area around a magnet where it exerts a force on a magnetic material 132

magnetic field lines the line in an area where magnetic effects are noticed; if you put a small compass on the line, the compass points along the line 132

magnetic materials something that will be attracted to a magnet 130

magnetic north the place on the Earth to which the north of a compass needle points 132

material substances from which objects are made 66

metamorphic rock formed from another type of rock as a result of heat and/or pressure 108

methane a gas, formula CH_4, produced by rotting materials and obtained from the ground when drilling for oil; also known as 'natural gas' 86

micro-organism a microscopic living thing; some cause disease 36

minerals (in biology) simple chemicals that plants and animals need to stay healthy 2, 4

minerals (in chemistry) the chemicals in rocks 96

mixture a substance in which two or more substances are mixed but not joined together 82

models objects or apparatus used to investigate or explain a phenomenon; in the mind, a group of ideas and pictures 9

molecule the smallest part of a chemical compound and the smallest part of an element that can exist in nature 68

moral concerned with the distinction between right and wrong 90

moral and ethical issues issues concerning whether things are right or wrong, or linked to any set of good principles 162

mouth start of the digestive system 10

N

neutralisation when an acid reacts with an alkali to make a neutral solution of a salt in water 80

nickel a metallic element that is attracted to magnets 130

noise pollution unwanted noise in the environment 160

non-luminous an object that does not produce light 144

non-magnetic materials materials that are not affected by magnets 130

non-vascular plants a group of plants without a specialised transport or vascular system 50

normal a line drawn at 90° to the point where a ray of light hits a mirror or a prism 146

north-seeking pole the end of a magnet that, if left to spin freely, will point towards the Earth's north pole 130

nutrients the food materials that cells use 2

O

object something which if placed in front of a mirror will form an image on the other side of the mirror 146

oesophagus the tube between your mouth and your stomach; also called the gullet 10

opaque will not allow light to pass through it 144

opinions what someone thinks, but not supported by conclusive evidence 60

oxygen a gas that makes up about one fifth of the air 18, 22

P

particles very small pieces of matter that everything is made of 68

penicillin antibiotic drug used to kill bacteria in the body 42

periodic table a table of the elements arranged in order so that similar elements are in the same column or group 72

periscope a device made from two mirrors, which can be used to see over walls or from submarines that are under water 146

phages a shortened name for bacteriophages 44

photosynthesis the process by which green plants convert carbon dioxide and water to food using sunlight; oxygen is produced as a result of the process 86

physical change a change that does not involve making any new substances; changes of state, dissolving and being broken in pieces are all types of physical change 70

physical weathering breakdown of rocks by physical processes such as water freezing and expanding in cracks, and by expansion and contraction of the surface as rocks heat up and cool down 98

plate boundaries the areas where tectonic plates meet; volcanoes and earthquakes tend to happen here 114

population all the plants or animals of one species that live in a particular place 54

porous describes the texture of a rock with pores 96

precipitation when a solid is formed in a solution 80

predictions statements about things we think will happen 74

preliminary tests tests, trial runs and information searches carried out to find out the best approach to an investigation 6

pressure how much pushing force there is on an area 108

primary sources original sources of information such as measurements in experiments; as opposed to secondary sources such as information from the Internet 138

prism a device, made of glass or plastic, which refracts light twice, it can be used to produce a spectrum 150

producer a name given to green plants because they produce food 58

properties how something behaves 72

protein nutrient needed for growth and repair; made up of amino acids 2, 8

pyramid of numbers pyramid-shaped diagram that shows how the numbers of living things change along a food chain 58

Q

quadrat an object, often a square frame, used for sampling living things 54, 60

qualitative described using words 138

quantitative described using numbers 138

R

radiation (of heat) a method of heat transfer in which the heat energy is given out as infra-red waves 122

range the difference between the lowest and the highest of a set of readings 104

ray a beam of light that travels in a straight line 142

reflected bounced back, for example light will reflect off a mirror 144

refraction bending of a ray of light when it travels from a material of one density to one of a different density 148

relay a device that uses an electromagnet to switch on one circuit when a second circuit is complete 136

repair mend damaged cells, tissues or organs 2

repel push apart 130

respiration the breakdown of food to release energy in living cells 18, 30

reverberations the repeated echoes of sound caused by reflections off the walls and surfaces in a room 158

risk factors the hazards of an activity or experiment 26

rock cycle the way that the material that rocks are made from is constantly moved around and changed from one type of rock to another 112

rock fragments tiny pieces of rock broken off by physical weathering 100

S

T

U

V

W

Acknowledgements

Alamy 8B.HSW.a (Blain Harrington III), 8D.2.b (David Hosking), 8E.2.a (Andrew Palmer), 8F.4.d (Phototake Inc.), 8K.HSW.a (Jupiterimages/Stock images); **Andrew Lambert** 8E.3.b, 8E.3c, 8E.4.a, 8F.4.b, 8F.5.b, 8G.1.a, 8G.1.b, 8G.1.c, 8G.1.d, 8G.1.e, 8G.1.f, 8G.1.g, 8G.2.b, 8G.2.c, 8G.2.d, 8G.4.d, 8G.HSW.a, 8L.1.b; **Cambridge University Press** 8G.4.a (Joanne Robinson); **Chris Westwood** 8G.3.c; **Corbis** 8B.3.a (Moonboard), 8B.3.b (Roger Ressmeyer), 8G.3.d (Michael Busselle), 8G.4.b (Chinch Gryniewicz/Ecoscene), 8H.1.h (Charles O'Rear), 8H.2.f (Yann Arthus-Betrand); **Ecoscene** 8D.3.a, 8D.3.b, 8D.3.d (Chinch Gryniewicz), 8D.3.h, 8D.3.k (Sally Morgan), 8D.3.j (Kevin King), 8F.6.b (Paul Thompson), 8F.HSW.c (Jim Winkley), 8F.HSW.e (Susan Cunningham); **education.co.uk/walmsley** 8L.1.c, 8L.3.c; **Fisher Scientific** 8I.HSW.a; **Geoscience Features Picture Library** 8F.2.c, 8F.5.a, 8G.2.f, 8H.1.a, 8H.1.b, 8H.1.c, 8H.1.e, 8H.1.f, 8H.1.g, 8H.2.b (Dr B.Booth), 8G.4.e (D. Bayliss), 8H.2.a (M. Hobbs), 8H.2.e (University of California); **Graham Burns** 8D.4.b; **Istituto e Museo di Storia della Scienza Florence** 8I.1.c; **Jean Martin** 8D.3.c, 8D.3.e, 8D.3.f; **Life File** 8F.6.a (Christopher Jones); **Mary Evans Picture Library** 8A.5.a; **Mediscan** 8A.5.c; **Nature Picture Library** 8L.3.a (Artur Tabor); **NHPA** 8D.HSW.a (Manfred Danegger), 8D.HSW.b (Roger Tidman), 8G.4.c (Daniel Heudin); **PA Photos** 8I.4.a (Phil Nobel/PA Archive); **Panos Pictures** 8F.HSW.d (Erik Schaffer); **Photolibrary** 8A.3.a, 8A.3.c (Martin Brigdale), 8A.3.b (Eaglemoss Consumer Publications); **Professional Sport** 8B.1.b, 8B.5.c (Tommy Hindley); **Redferns Music Picture Library** 8L.1.a; **Robert Harding Picture Library Ltd** 8F.3.b (S. Frieberg), 8G.2.e (John Start); **Science Photo Library** 8A.1.b (Peter Menzel), 8A.1.c, 8A.1.d (Biophoto Associated), 8A.5.b (Alain Pol, ISM), 8B.1.a (Cristina Pedrazzini), 8B.1.c (Jeremy Walker), 8B.2.a (Sheila Terry), 8B.2.b (Alfred Pasieka), 8B.3.c (James King-Holmes), 8B.4.a (Alfred Pasieka), 8C.1.a (R. Maisonneuve, Publiphoto Diffusion), 8C.2.a (Jane Shemitt), 8C.4.a (Noble Proctor), 8C.HSW.a (Louise Murray), 8C.HSW.b (Volker Steger), 8D.1.a (Dr. Jeremy Burgess), 8D.3.g (John Heseltine), 8D.2.a (Vanessa Vick), 8D.2.c (Duncan Shaw), 8D.2.d, 8D.4.a (Lepus), 8E.1.b (Pascal Goetgheluck), 8E.1.c 9Claude Nuridsany & Marie Perennou), 8E.2.b (Susumu Nishinaga), 8E.3.d (Jerry Mason), 8E.HSW.a, 8F.1.a (Alfred Pasieka), 8F.1.b (Andrew McClenaghan), 8F.2.a (Martyn F. Chillmaid), 8F.2.b (Charles D. Winters), 8F.3.a (Sheila Terry), 8F.4.a (Tony McConnell), 8F.4.c (David Taylor), 8F.HSW.a (J.C. Revy), 8F.HSW.b (Biophoto Associates), 8F.HSW.f (Martin Bond), 8G.2.g (Tony Craddock), 8G.3.e (John Mead), 8H.1.d (George Bernard), 8H.1.i (G. Brad Lewis), 8H.1.j (Martin Bond), 8H.2.c (Oscar Burriel), 8H.2.d (Soames Summerhays), 8I.1.a (Astrid & Hans Frieder Michler), 8I.1.b (Chris Priest & Mark Clarke), 8K.HSW.b (R. Maisonneuve, Publiphoto Diffusion), 8L.HSW.a (Dr Najeeb Layyous), 8L.HSW.b (Saturn Stills); **The Allan Cash Picture Library** 8E.3.a; **Vanessa Miles** 8B.5.a, 8B.5.b, 8D.3.i, 8E.1.a, 8G.2.a, 8G.3.b, 8J.1.a, 8L.3.b; **Wellcome Photo Library** 8A.1.a (Fiona Progoff), 8A.2.a; **Wilderness Photographic Library** 8G.3.a.

Image references show the Unit and Topic of the book (eg. 8A.1) and the order of the image in the Topic from top to bottom, left to right (e.g. 8A.1.b is the second photograph in Topic 1 of Unit 8A).

Series advisors	Andy Cooke, Jean Martin
Series authors	Sam Ellis, Jean Martin
Series consultants	Diane Fellowes-Freeman, Richard Needham

Based on original material by Derek Baron, Trevor Bavage, Paul Butler, Andy Cooke, Zoe Crompton, Sam Ellis, Kevin Frobisher, Jean Martin, Mick Mulligan, Chris Ram